Global Issues

WITHDRAWN

To my parents
Grace and George Seitz
and to my children
Leksi and Abigail

GLOBAL ISSUES:
An Introduction

John L. Seitz

BLACKWELL
Publishers

First published 1995

Reprinted 1996, 1997, 1998

Blackwell Publishers Inc
350 Main Street
Malden, Massachusetts 02148, USA

Blackwell Publishers Ltd
108 Cowley Road
Oxford OX4 1JF, UK

Library of Congress Cataloging in Publication Data
Seitz, John L., 1931–
Global issues: an introduction/John L. Seitz.
p.   cm.
Includes bibliographical references and index.
ISBN 1–55786–439–X (hb) — ISBN 1–55786–440–3 (pb)
1. Economic development. 2. Developing countries — Economic policy.
3. Developing countries — Economic conditions. 4. Economic history — 1945–
I. Title.
HD82.S416   1995   95–14675
338.9 — dc20   CIP

British Library Cataloguing in Publication Data
A CIP catalogue record for this book is available from the British Library

Production Manager/Controller: Emma Gotch
Text Designer: Ian Foulis & Associates
Picture Designer: Ian Foulis & Associates

Typeset in Caslon 224 10.5pt on 12pt
by Ian Foulis & Associates, Saltash, Cornwall
Printed and bound in Great Britain by T. J. Press Ltd, Padstow, Cornwall

This book is printed on acid-free paper

CONTENTS

5   The Environment   138

List of Plates

ix

List of Figures and Tables

Introduction

In the 1950s and 1960s I went as an employee of the US government to Iran, Brazil, Liberia, and Pakistan to help them develop. A common belief in those decades was that poverty causes people to turn to communism. As an idealistic young person, I was pleased to work in a program that had the objective not only of controlling the spread of communism but also of helping poor nations raise their living standards. After World War II the United States was the richest and most powerful country in the world. Many countries welcomed US assistance since it was widely believed that the United States could show others how to escape from poverty.

Disillusionment came as I realized that we did not really know how to help these countries relieve their widespread poverty. The problem was much more complex and difficult than we had imagined. Also, one of the main political objectives of our foreign aid program – to help friendly, noncommunist governments stay in power – often dominated our concerns.

And more disillusionment came when I looked at my own country and realized that it had many problems of its own that had not been solved. It was called "developed" but faced major problems that had accompanied its industrialization – urban sprawl and squalor, pollution, crime, materialism, and ugliness, among others. So, I asked myself, What is development? Is it good or bad? If there are good features in it, as many people in the world believe, how do you achieve them, and how do you control or prevent the harmful features? It was questions such as these that led me to a deeper study of development and to the writing of this book.

I came to recognize that development is a concept that allows us to examine and make some sense out of the complex issues the world faces today. Many of these issues are increasingly seen as being global issues. Because the capacity human beings have to change the world – for better or for worse – is constantly growing, an understanding of global issues has become essential. The front pages of our newspapers and the evening TV news programs remind us nearly daily that we live in an age of increasing interdependence. (The Foreword explains the creation of global issues.)

In this book the term "development" will be defined as economic growth plus the social changes caused by or accompanying that economic growth. In the 1950s and 1960s it was common to think of development only in economic terms. For many economists, political scientists, and government officials, development meant an increase in the per capita national income of a country or an increase in its gross national product (GNP), the total amount of goods and services produced. Development and economic development were considered to be synonymous. In the 1970s an awareness grew – in both the less developed[1] nations of the Third World[2] and the developed industrialized nations – that some of the social changes generated by economic growth were undesirable. More people were coming to understand that for economic development to result in happier human beings, attention would have to be paid to the effects that economic growth was having on social factors. Were an adequate number of satisfying and challenging jobs being created? Were adequate housing, health care, and education available? Were people living and working in a healthy and pleasant environment? Did people have enough nutritious food to eat? Every country is deficient in some of these factors and, thus, is in the process of developing.

The definition of development I have given above is a "neutral" one – it does not convey a sense of good or bad, of what is desirable or undesirable. I have chosen this definition because there is no widespread agreement on what these desirable and undesirable features are. The United Nations now defines human development as the widening of human choices, and in a yearly publication ranks nations on a human development index, which tries to measure national differences of real purchasing power, education, and health.[3] Economists have traditionally used GNP or national income as the measures of economic development. My definition tries to combine both the economic and the social components into the concept of development. I find a neutral definition useful because development can be beneficial or harmful to people.[4]

In this book we will look at some of the most important current issues related to development. The well-being of people depends on how governments and individuals deal with these issues. We will first look at the issue of poverty in the world and then move on to issues related to population, food, energy, the environment, and technology, and we will conclude with a consideration of the future.

This book is an introduction to a number of complicated issues. It is only a beginning; there is much more to learn. Readers who are intrigued by a subject or point made and want to learn more about it should consult the relevant footnote, which will either give some additional information or will give the source of the fact I present. Consulting this source is a good place for the reader to start his or her investigation. After each chapter a list of readings gives inquisitive readers further suggestions for articles and books that will allow them to probe more deeply. The selected bibliography contains some additional books and articles. Appendix 1 gives the student some help in organizing the material the book covers and the teacher some suggestions for teaching this material. Appendix 2 offers suggestions of relevant videotapes, an important and interesting resource for those who want to understand these issues more deeply. Appendix 3 gives study and discussion questions for each chapter for use by students and teachers.

This book would not have been written if I had not received a grant from the Andrew W. Mellon Foundation. This grant, which was administered by Wofford College, allowed me to be excused from my teaching duties long enough to get the project well underway. Numerous colleagues have assisted me with their comments on parts of the manuscript. Especially useful were comments from Ab Abercrombie, Bud Talley, Crawford Young, Tom Oleszczuk, who read all or most of the original manuscript; Edwin Clausen, Eugene Shultz, Thomas O'Toole, Milton Krieger, Rickie Sanders, and Susan Farnsworth read the revised version. Valuable secretarial services were provided in a superb manner by Joyce Blackwell. Jeanne Cheatham put most of the chapters of the original manuscript on computer disks, which greatly facilitated the preparation of the book. Karen Plaszaj, Wendy Sellers, and Lance Crick provided research assistance.

I extend thanks also to Linton Dunson, the head of my department at Wofford for many years, whose management philosophy of "hire the best people you can afford and then leave them alone" helped to create an atmosphere that supported creativity.

My deep appreciation goes to John Davey, Editorial Director at Blackwell Publishers, who "discovered" the book and who, along with Jane Robertson, Blackwell's Managing Editor, made numerous suggestions that strengthened the book.

Male authors occasionally conclude with a word of thanks to their wives for having typed the manuscript or kept the children quiet. Mine didn't do either. But my wife, Merike Tamm, gave me the idea for the course from which this book evolved. She has also been an excellent editor and has given loving support while the work progressed. For this I thank her.

NOTES

1 The term "less developed" refers to a relatively poor nation in which agriculture or mineral resources are dominant in the economy while manufacturing and services play a minor role. The infrastructure (transportation, education, health, and other social services) of these countries is usually inadequate for their needs. About 80 percent of the world's people live in nations such as this, which are also called "developing." (Some of these countries are highly developed in culture, and many such regions of the world had ancient civilizations with architecture, religion, and philosophy that we still admire.) Since many of the less (economically) developed nations are in the Southern

Hemisphere, they are at times referred to as "the South." Industrialized nations are called "developed" nations. Most of them are located in the Northern Hemisphere so they are called "the North."

2 "Third World" refers to the poorer nations, or developing nations, of Latin America, Asia, and Africa. "First World" used to refer to capitalist industrial nations, while "Second World" referred to communist industrial nations. "Fourth World" has sometimes been used to refer to the poorest part of the Third World with little chance for becoming industrialized. Although "Third World" is still commonly used, it is a term that is increasingly seen to be out-of-date. The terms "First World" and "Second World" have not been used much since the collapse of the Soviet Union. Also, the economic and social differences among the nations in the Third World group is huge, which limits the usefulness of the term. I will continue to use Third World in this book because an adequate replacement has not yet been created, although "South" is increasingly being used in its place.

3 United Nations Development Programme, *Human Development Report 1993* (New York: Oxford University Press, 1993) , p. 10. The organization Population Action International of Washington, DC, publishes a "human suffering index" every five years that rates countries according to ten measures of human welfare: income, inflation, demand for new jobs, urban population pressures, infant mortality, nutrition, clean water, energy use, adult literacy, and personal freedom.

4 For a criticism of the Western concept of development, see Ivan Illich, "Outwitting the 'Developed' Countries," in Charles K. Wilber (ed.), *The Political Economy of Development and Underdevelopment,* 2nd edn (New York: Random House, 1979), pp. 436–44. See also, Lloyd Timberlake, "The Dangers of 'Development'," in *Only One Earth: Living for the Future* (New York: Sterling Publishing, 1987), pp. 13–22.

Foreword:

The Creation of Global Issues

What causes an issue to become a "global issue"? Are "global issues" the same as international affairs – the interactions that nations, organizations, and peoples from different countries have with each other? Or is something new happening in the world? Are there now concerns and issues that are increasingly being recognized as global in nature? It is the thesis of this book that something new indeed is happening in the world as nations become more interdependent. While their well-being is still largely dependent upon how they run their internal affairs, increasingly nations are facing issues that they alone cannot solve, issues that are so important that the failure to solve them will adversely affect the lives of many people on this planet. In fact, some of these issues are so important that they can affect how suitable this planet will be in the future for supporting life. What factors are involved in the creation of a global issue?

First of all, some event, or a series of events in the news, catches the world's attention. Maybe it is a cataclysmic event, such as the explosion of the nuclear plant in Chernobyl, which dramatizes the issue of the safety of nuclear power. Or maybe it is a series of important events – such as the Iranian revolution, OPEC price increases, and the Iraqi invasion of Kuwait – which convinces many people that there is an energy crisis. Or maybe there is a series of announcements by scientists, such as the announcements that the ozone layer is being depleted. What all these news items have in common is a that they were recognized as having the potential of inflicting serious harm on many people in many nations.

A second factor needed for the creation of a global issue is that the problem does not go away. News of it may surface and recede, but the problem remains. Often it is a problem that will not be easy to solve, or that will grow worse without efforts to attack it. The population explosion in the less developed nations is a clear example of such a problem.

Third, the problem seems to dramatize our increasing interdependence. The communications and transportation revolutions that we are

1

experiencing are giving people knowledge of many new parts of the globe. We see that what is happening in far-off places can affect, or is affecting, our lives. Instability in the oil-rich Middle East affects the price of oil around the world and since many countries are dependent on oil as their main source of energy, the politics of oil becomes a global concern.

Fourth, our increasing knowledge of the world also makes some issues impossible to forget. Although some of these issues, such as hunger in Africa, might not directly affect many of us, our new knowledge of them – powerfully aided by pictures (television and print) – makes the issue a global one. A part of the human race is suffering in ways no humans should have to suffer. The conscience of those more fortunate can be awakened by this new knowledge, and these people might now recognize the need for action.

And fifth, global issues are often seen as being interrelated. One issue affects other issues. For example, global warming (an environmental issue) is related to an energy issue (our reliance on fossil fuels), the population issue (more people produce more greenhouse gases), the wealth and poverty issue (wealthy developed countries produce the most gases that cause the warming), and the technology issue (new technology can reduce some of the greenhouse gases).

Perhaps, global issues were born on the day, several decades ago, when the earth, for the first time, had its picture taken. The first photograph of earth, which was transmitted by a spacecraft, showed our planet surrounded by a sea of blackness. Many people seeing that photograph realized that the blackness was a hostile environment, devoid of life, and that life on earth was vulnerable and precious. No national boundaries could be seen from space. That photograph showed us our home – one world – and called for us to have a global perspective in addition to our natural, and desirable, more local and national perspectives.

This book discusses *some* of the main global issues of our time. The reader can probably identify others. During the reader's lifetime, humanity will have to face new global issues that will continue to surface. It is a characteristic of the world in which we live. Maybe our growing ability to identify such issues, and our increasing knowledge of how to deal with them, will enable us to handle the new issues better than we are doing with the present ones.

1

The Wealth and Poverty of Nations

T he mere fact that opposing visions of economic development have grown to shape the international agenda is in one sense merely an indication that development concerns are receiving attention on a global scale for the first time in history.

Lynn H. Miller, *Global Order: Values and Power in International Politics* (1985)

For most of history, human beings have been poor. A few individuals in many societies had a higher standard of living than their fellow humans, but the vast majority of people on earth shared a common condition of poverty. The Industrial Revolution brought a fundamental change. New wealth was created in the industrializing nations in Europe and eventually shared by larger numbers of people. And the differences between the rich and the poor in the world began to increase. A few nations began to achieve higher living standards, and they began to pull away from the rest of the world, which had not yet begun to industrialize. It is estimated that around 1850 the difference between the average incomes of people in industrializing Europe and in nonindustrial countries was 2 to 1. By 1950, the income gap between rich and poor countries was estimated to have grown to 10 to 1; in 1960 it was 15 to 1. If present trends continue, it is now predicted that the gap between rich and poor nations could reach 30 to 1 by the end of the century.[1] The United Nations expresses this trend in another way: the gap between the income of the richest 20 percent of the world's people and the poorest 20 percent doubled during the past 30 years. In the early 1990s the richest 20 percent were receiving 150 times the income of the poorest 20 percent.[2] In other words, the rich earned $1,500 for every $10 the poor earned.

Poverty in Indonesia (*World Bank*)

The growing gap between the rich and the poor is only one part of the picture of world-wide economic conditions. Another important development is the improvement in living standards for many, even in the poorer nations. While it is certainly true that the benefits of the impressive economic growth that has occurred in many developing nations since the end of World War II have been very unevenly shared – with many of the poorest receiving very little benefit, if any at all – it is also true that the lives of many others have improved. In a single 25-year period, from the late l960s to the early 1990s, the average consumption per capita in developing countries increased by 70 percent in real terms, average life expectancy rose from about 50 to 65 years, and primary school enrollment rates reached nearly 90 percent.[3] Table 1.1 shows the growth of real per capita income in the industrial and the developing countries from 1960 to 1990. An improvement was made in income in the developing countries during this period, although conditions worsened in the 1980s in three regions: sub-Saharan Africa, the Middle East and North

Africa, and Latin America. (Averages, such as those shown in the table, do not reveal the vast differences in incomes that exist in many countries, especially in many developing countries where the income distribution is highly unequal. In most nations 60 to 70 percent of the people earn less than the nation's average income.[4])

Table 1.1 Growth of real per capita income[a] in industrial and developing countries, 1960–90

	1960–70	1970–80	1980–90
Industrial countires	4	2	2
Developing countires	3	3	1
South Asia	1	1	3
East Asia	4	5	6
Sub-Saharan Africa	1	1	−1
Middle East and North Africa	6	3	−3
Latin America	3	3	−1

[a]average annual percentage change

Source: Adapted from World Bank, *World Development Report 1992 - Development and the Environment* New York: Oxford University Press, 1992), p. 32.

The 1980s have been called a "lost decade" for the poor of the world, and as both tables 1.1 and 1.2 show, this was true for large numbers of the poor in Latin America, sub-Saharan Africa, and the Middle East. The Persian Gulf War and declining oil revenues played a significant role in the worsening conditions in the Middle East, while high population growth, natural disasters, and civil wars generated them in Africa. Many Latin American and African countries were also plagued by falling prices for their export commodities, large interest payments on huge debts owed to foreign banks and governments, and poor economic policies.

The hope that was fairly common in the 1960s of reducing the gap between the rich and the poor in the world had all but disappeared by the 1980s. Today, the world continues to endure vast inequalities of wealth. As can be seen in table 1.2, in 1990 about 30 percent of the people in the less developed nations had annual incomes below the poverty line, which the World Bank set at about $400. While the number of poor people decreased in East and South Asia from 1985 to 1990, in sub-Saharan Africa, the Middle East and North Africa, and Latin America the number living below the poverty line actually increased, a trend that is expected to continue through the year 2000. In 1990 there were about 1.1 billion poor people in the world, about one fifth of the world's population.

Who are the poor? According to the World Bank, nearly half of them live in South Asia (e.g., India, Pakistan, Bangladesh), while a smaller but highly disproportionate number live in Africa, south of the Sahara desert. Within regions and countries the poor tend to be concentrated in rural areas with a high density of population, such as on the Gangetic

Plain in India and on the island of Java in Indonesia. Many poor also live in areas with scarce resources such as in the Andean highlands in Latin America and in the Sahel region in Africa.[5] The weight of poverty in the less developed nations falls heaviest on women, children, and minority ethnic groups.

Table 1.2 Poverty in the developing world, 1985–2000

	Percentage of population below the poverty line[a]			Number of poor (millions)		
	1985	1990	2000 (projected)	1985	1990	2000 (projected)
All developing countries	31	30	24	1,050	1,130	1,110
South Asia	52	49	37	530	560	510
East Asia	13	11	4	180	170	70
Sub-Saharan Africa	48	48	50	180	220	300
Middle East and North Africa	31	33	31	60	70	90
Latin America	23	26	25	90	110	130

[a] In 1990 prices, the poverty line was about $400 annual income per capita

Source: Adapted from World Bank, *World Development Report, 1992 - Development and the Environment* (New York: Oxford University Press, 1992), p. 30.

Consequences of poverty: the mothers who don't cry

In trying to understand what being poor means in a less developed country, a puzzling question comes to mind: Why don't many mothers cry when their children die in northeastern Brazil?

To answer this question, one needs to know first why the death of a child is fairly common in the Brazilian northeast, the poorest region in Brazil, and one of the poorest in the world. A North American anthropologist offers this explanation:

The children of the Northeast, especially those born in shantytowns on the periphery of urban life, are at a very high risk of death. In these areas, children are born without the traditional protection of breast-feeding, subsistence gardens, stable marriages, and multiple adult caretakers that exists in the interior [of Brazil]. In the hillside shantytowns that spring up

around cities... marriages are brittle, single parenting is the norm, and women are frequently forced into the shadow economy of domestic work in the homes of the rich or into unprotected and oftentimes "scab" wage labor on the surrounding sugar plantations, where they clear land for planting and weed for a pittance, sometimes less than a dollar a day. The women... may not bring their babies with them into the homes of the wealthy, where the often sick infants are considered sources of contamination, and they cannot carry the little ones to the riverbanks where they wash clothes because the river is heavily infested with schistosomes and other deadly parasites... At wages of a dollar a day, the women... cannot hire baby sitters. Older children who are not in school will

The weight of poverty falls heavily on children in less developed nations. *(United Nations)*

sometimes serve as somewhat indifferent caretakers. But any child not in school is also expected to find wage work. In most cases, babies are simply left at home alone, the door securely fastened. And so many also die alone and unattended.

The death of a child is thus a commonplace tragedy in northeast Brazil. But why is there a lack of mourning for the dead children? The anthropologist found that no tears were shed when an infant died and that mothers never visited the graves after the burials. The anthropol-ogist concluded that in the face of the frequent deaths of children, mothers have learned to delay their attachment to any child until that child has proved to be a survivor, hardier than his or her weaker siblings. This reaction to the realities of their lives has allowed these women to continue living in a harsh situation and to not let grief make their lives unbearable. Actions similar to those of the Brazilian women have been observed in parts of Africa, India, and Central America.

Source: Nancy Scheper-Hughes, "Death Without Weeping," *Natural History* (October 1989), pp. 8 and 14.

Consequences of poverty: the children who are killed

Another question also illustrates one of the many consequences of poverty. Why are children being killed in some of the cities and rural areas of Brazil, some even by the police? The answer is that children are often involved in begging or crime. They become the targets of drug dealers who are enforcing control of their gangs, or they are killed by assassins, some of them off-duty policemen who are hired by merchants or residents trying to control crime in their areas. At least 5,000 children were killed between 1988 and 1990. Some of these victims had been abandoned by their parents while others were living with single parents or friends, many in the large shantytowns surrounding many Brazilian cities. And there are other depressing statistics: In 1992 about 1,000 children were sleeping on the streets of Rio de Janeiro alone; three-quarters of Brazil's children do not finish primary school. Poverty inflicts a harsh life for many children in other countries also. In Nairobi, Kenya, it has been estimated that about 50,000 children roam the streets, many because their single parents cannot afford the fees at government schools. Forty-four *million* children reportedly work on the streets in India; 10 million children work on the streets in Mexico and about 250,000 actually live on the streets; and in Thailand an estimated 800,000 girls work on the streets or in brothels as prostitutes (some girls being forced into prostitution are as young as eight years old).

Sources: Jane Perlez, "Nairobi Street Children Play Games of Despair," *New York Times*, national edn (January 2, 1991), p. A1; James Brooke, " Gunmen in Police Uniform Kill Seven Street Children in Brazil," *New York Times*, national edn (July 24, 1993), pp. 1 and 3; Werner Fornos, "Children of the Streets: A Global Tragedy," *Popline* (Washington: Population Institute, November-December 1991), p. 3. The increasing use of children as prostitutes because of family poverty and the fear of AIDS is reported in Marlise Simons, "The Sex Market: Scourge on the World's Children," *New York Times*, national edn (April 9, 1993), p. A3.

Street children in Nepal.
(*Ab Abercrombie*)

Now that we have made a brief examination of poverty in the world, let's now focus another question: Why are some countries rich and some poor? There is no agreement on the answer to that question, but various views have been presented over the years. Although vast differences among the nations of the world make generalizations hazardous, it can be useful to consider three of the most widely accepted approaches or views of economic development: (1) the market or capitalist approach, which is adhered to by many in the Western industrial countries;[6] (2) the state approach, varieties of which were commonly believed in in communist countries and, in the past, were believed in by many in the Third World; and (3) the civil society approach.[7]

The Market Approach ●

The market approach holds that nations can acquire wealth by following four basic rules: (1) the means of production – those things required to produce goods and services such as labor, natural resources, technology, and capital (buildings, machinery, and money that can be used to purchase these) – must be owned and controlled by private individuals or firms; (2) markets in which the means of production and the goods and services produced are freely bought and sold must exist; (3) trade at the local, national, and international levels must be unrestricted; and (4) a state-enforced system of law must exist to guarantee business contracts so as to make safe commercial relations between unrelated individuals.

Adam Smith, the eighteenth-century Scottish political economist and founder of the market approach, believed that the operations of labor are the key to increasing production. He argued that it is much more efficient for workers to specialize in their work, focusing on one product rather than making many different products. If workers do this, and if they are brought together in one location so their labor can be supervised, increased production will result. Smith also presented the idea that, if the owners of the means of production are allowed to freely sell their services or goods at the most advantageous price they can obtain, the largest amount of products and services will be produced and everyone will benefit. It is the prices in the markets that suggest to the businessman or businesswoman new profitable investment opportunities and more efficient production processes. (For example, when oil prices rose dramatically in the 1970s, new investments occurred in alternative energy sources and some industries came up with ways to reduce the amount of oil they needed to buy. Some business people saw the alternative energy investments as a way for them to make money in the energy field, and some industries cut their costs, thus increasing their profits, by becoming more efficient in their use of energy.)

Smith did not focus on the role of the entrepreneur, but later market theorists did, making the entrepreneur – the one who brought the means of production together in a way to produce goods and services – a key component in this approach. Finally, Smith and other market theorists emphasized the importance of free trade. If a nation concentrates on producing those products in which it has a comparative advantage over other nations, advantages that climate, natural resources, cheap labor, or technology give it, and if it trades with other nations that are also concentrating on those products that *they* have the greatest advantage in producing, then all will benefit.

The market approach holds that government has a crucial but limited role in maintaining an environment in which economic activities can flourish. Government should confine its activities to providing for domestic tranquillity that would ensure that private property is protected; providing certain services, such as national defense, for which everyone should pay; enforcing private contracts; and helping to maintain a stable supply of money and credit. The reason some nations are poor, according to the market approach, is that they do not follow these basic rules.

Advocates of the market approach point to the wealth of the United States and Western Europe as evidence of the correctness of their view. Even Karl Marx said that the 100 years of rule by capitalists were the most productive in the history of the world. And although an uneven distribution of income occurred in Western Europe during its early period of industrialization, the distribution of income later became much less uneven. This indicated that the new wealth was being shared by more and more people.

Nations such as Japan and West Germany, which came back from the devastation of World War II to create extremely strong economies by following the basic principles of the market approach, are also cited as evidence of the validity of the approach. Examples can also be found among developing nations that have achieved such impressive economic growth by following the principles of this approach that they have moved into a separate category of the Third World: the newly industrializing states. Many of these states, such as South Korea, Taiwan, Singapore, and the British colony Hong Kong, achieved their high economic growth mainly by exporting light manufactured products to the developed nations.

Finally, advocates of the market approach point to the decisions of many ex-communist and developing countries during the 1980s to adopt at least some market mechanisms in their efforts to reform their economies. Even China – the largest remaining communist government – has adopted many important aspects of the market approach; it is this adoption that is widely believed to be responsible for China's impressive economic growth.

Critics of the market approach point to the high rates of unemployment that have existed at times in Western Europe and the United States. At the present time, high unemployment exists throughout the Third World, even in a number of nations that follow the market

The market approach is followed on the streets of many developing nations. *(Mark Olencki)*

approach and had impressive increases in their GNP. Much of the industry that has come to the Third World has been capital intensive; that is, it uses large amounts of financial and physical capital but employs relatively few workers.

There is evidence that in countries such as Brazil, which has basically followed the market approach for the past several decades, the distribution of income within the countries became more unequal during the period the countries were experiencing high rates of growth. The rich got a larger proportion of the total income produced in the countries than they had before the growth began. And even worse than this is the evidence that the poor in countries such as Brazil probably became absolutely poorer during the period of high growth, in part because of the high inflation that often accompanied the growth.[8] (High inflation usually hurts the poor more than the rich because the poor are least able to increase their income to cope with the rising prices of goods.) The economic growth that came to some developing nations following the market approach thus failed to trickle down to the poor and, in fact, may have made their lives worse.

Critics of the market approach have also pointed out that prices for goods and services set by a free market often do not reflect the true costs of producing those goods and services. Damage to the environment or to people's health that occurs in the production and disposal of a product is often a hidden cost not covered by the price of the product. The market treats the atmosphere, oceans, rivers, and lakes as "free goods," and, unless prohibited from doing so by the state, it transfers the costs that arise because of their pollution to the broader community. Take for example a factory that pollutes the air while producing cars; the costs of treating illnesses caused by damaged lungs are not borne by the factory owner or the purchaser of the car, but rather by the community as a whole.

The State Approach

The state approach, which was founded mainly by Karl Marx, German philosopher and political economist, and Vladimir Ilyich Lenin, Russian revolutionary leader and first premier of the Soviet Union, has more to say about the causes of underdevelopment than it does about how development takes place. In a socialist country most of the means of production – land, resources, and capital – are publicly controlled to ensure that the profit obtained from the production of goods and services is plowed back to benefit the community as a whole. The prohibition on the private control or ownership of these so-called factors of production leads, according to the state approach, to a relatively equal distribution of income as everyone, not just a few individuals, benefits from the economic activity. The basic needs of all are provided for. The free market of the capitalist system is abolished and replaced with central planning.

12

Prices are set by the central planners, and capital is invested in areas that are needed to benefit the society.

The explantion the state approach gives to the causes of poverty in the world focuses on international trade. According to the state approach, the root of the present international economic system where a few nations are rich and the majority of nations remain poor lies in the trade patterns developed in the sixteenth century by Western Europe.[9] ("Dependency theory" is the name given to this part of the state approach.) First Spain and Portugal and then Great Britain, Holland and France gained colonies – many of them in the Southern Hemisphere – to trade with. The imperialistic European nations in the Northern Hemisphere developed a trade pattern that we can still see clear signs of today. The mother countries in "the core" became the manufacturing and commercial centers, and their colonies in "the periphery" became the suppliers of food and minerals. Railroads were built in the colonies to connect the plantations and mines to the ports. This transportation system, along with the discouragement of local manufacturing competing with that done in the mother countries, prevented the economic development of the colonies. The terms of trade – what one can obtain from one's exports – favored the European nations, since the prices of the primary products from the colonies remained low while the prices of the manufactured products sent back to the colonies continually increased.

When most of the colonies gained their independence after World War II, this trade pattern continued. Many of the less developed countries still produce food and minerals for the world market and primarily trade with their former colonial masters. The world demand for the products from the poorer nations fluctuates greatly, and the prices of these products remain depressed. The political and social systems that developed in the former colonies also serve to keep the majority within these developing nations poor. A local elite, which grew up when these countries were under colonial domination, learned to benefit from the domination by the Western countries. In a sense, two societies were created in these countries: one, relatively modern and prosperous, revolved around the export sector, while the other consisted of the rest of the people who remained in the traditional system and were poor. The local elite, which became the governing elite upon independence, acquired a taste for Western products, which the industrial nations were happy to sell them at a good price.

The present vehicle of this economic domination by the North of the South is the multinational corporation. Over 10,000 of these exist today, most with headquarters in the United States but a growing number of others headquartered in Europe and Japan. These corporations squeeze out small competing firms in the developing nations, evade local taxes through numerous devices, send large profits back to their headquarters, and create relatively few jobs since the manufacturing firms they set up utilize the same capital-intensive technology that is common in the industrialized countries. Also, they advertise their products extensively, thus creating demands for things such as Coca Cola and color television sets while many people in the countries in which they operate still do not have enough to eat.

The advocates of the state approach point to the adverse terms of trade that many developing nations face today. There is general agreement that there has been a long-term decline in the terms of trade for many agricultural and mineral products that the less developed nations export. There has also been great volatility in the prices of some of these products, with a change of 25 percent or more from one year to the next not uncommon for cocoa, rubber, sugar, copper, lead and zinc.[10] Such fluctuations make economic planning by the developing nations very difficult. There is also clear evidence that the industrialized countries, while primarily trading among themselves, are highly dependent on the less developed countries for many crucial raw materials, including chromium, manganese, cobalt, bauxite, tin, and, of course, oil.

Although international trade is still far from being the most important component of the US economy, it is a very important factor for many of the wealthiest corporations. In the early 1980s about one-half of the 500 wealthiest corporations listed in *Fortune* magazine obtained over 40 percent of their profits from their foreign operations.[11] And some multinational corporations have financial resources larger than those of many Third World nations.

Finally, the defenders of the state approach argue that there is little chance for many poor nations to achieve a distribution of income as fair as that achieved by Europe after it industrialized. This situation has evolved because the controlling elites in underdeveloped nations today have repressive tools at their disposal (such as sophisticated police surveillance devices and powerful weapons) that the European elites did not have. This allows them to deal with pressures from the "have-nots" in a way that the Europeans never resorted to.

Critics of the state approach point to the breakup of the Soviet empire in Eastern Europe in the late 1980s and to the collapse of communism in the former Soviet Union and the breakup of that country in the early 1990s as support for their view that the state approach cannot efficiently produce wealth. In fact, it was the dissatisfaction of Eastern Europeans with their economic conditions that played a large role in their massive opposition to the existing communist governments and their eventual overthrow. Dissatisfaction with economic conditions also played a large role in the overthrow of the Soviet government, a startling rejection of the state approach by a people who had lived under it for 70 years.

Critics of the state approach also point to the suppression of individual liberties in the former Soviet Union, China, and other communist states as evidence that the socialist model for development has costs that many people are not willing to pay. In fact, most revolutions have huge costs, leading to much suffering and economic deterioration before any improvement in conditions is seen; even after improvements occur,

The state approach to development struggles to survive the collapse of communist regimes in Europe as can be seen in the posters of a Communist Party conference in Nepal. *(Ab Abercrombie)*

oppressive political and social controls are used by leaders to maintain power.

Central planning has proved to be an inefficient allocator of resources wherever it has been followed. Without prices from the free market to indicate the real costs of goods and services, the central planners cannot make good decisions. And if efficient central planning has proved to be impossible in a developed country such as the former Soviet Union, it has proved to be even worse in underdeveloped nations where governmental administrative capability is weak. A final negative feature of central planning is that it always leads to a large governmental bureaucracy.

Multinational corporations have created jobs in the Third World that would not have existed otherwise; they have brought new technologies to the less developed nations and they have helped those nations' balance of payments problems by bringing in scarce capital and by helping develop export industries that earn needed foreign exchange. These advantages are well known in the Third World, and explain why multinational corporations are welcomed by many less developed countries.

Finally, the critics of the state approach argue that political elites in developing nations have used the dependency theory, especially in Latin America where the theory is popular, to gain local political support among the bureaucracy, the military, and the masses. To blame the

15

industrial nations for their poverty frees them from taking responsibility for their own development and excuses their lack of progress. It also frees them from having to clean their own houses of governmental corruption and incompetence, and to stop following misguided economic development approaches. The newly industrializing countries have shown that when market principles are followed, economic progress can be made even by developing nations that have a dense population and few, if any, natural resources.

The Civil Society Approach

In the words of a Harvard anthropologist, civil society is "the space between the state and the individual where those habits of the heart flourish that socialize the individual and humanize the state."[12] In more simple terms, civil society is the activity that people engage in as they interact with other people, and it can be seen in families, neighborhoods, voluntary organizations, and spontaneous grassroots movements. Although this activity can be directed toward economic gain, often it is not. It is the activity that makes a community, a connection between people, a realization that each individual is dependent upon others and that they share life together. Without civil society, isolation results and since human beings are social animals, that isolation can lead to illness, antisocial behavior, and even suicide.

The civil society approach to development emphasizes social development, how people act toward other people. But the approach can also have important political and economic aspects. The best way to demonstrate this is through examples. In 1973 a group of poor people in India rushed to the forests above their impoverished village and hugged the trees to prevent a timber company from cutting them down. This community action received world-wide publicity and helped to force some governments to reconsider their development policy regarding their nation's forests. The Chipko movement, which grew out of this action, is an example of self-help community action directed against threats and harm to the environment, harm that the local people realize will make their lives more difficult or even impossible.

Civil society can also be seen functioning in the efforts by some people in poorer countries to raise their low living standards. It is generated by the realization in many Third World countries that neither their governments nor the market can be relied on to help their citizens obtain basic needs. Here are two examples: In Latin America after the bishops of the Catholic Church met in 1968 in Colombia and decided that the Church should become active in helping the poor, many priests, nuns, and lay Christians helped form Christian Base Communities, self-help groups mainly made up of the poor themselves. In Bangladesh a professor concluded that the landless poor could never improve their conditions without some extra funds to help them start up an income-producing activity. Since no banks would lend them money, he set up the Grameen Bank. The bank's loans, averaging $60 each, were repaid at a much higher rate than loans at regular banks: the Grameen Bank achieved a 95 percent repayment rate![13] This experience demonstrated that the poor can be good financial risks and has been imitated in other parts of the world.

Civil society can also be directed toward political goals. In Eastern Europe in the 1980s millions of citizens took to the streets to call for the end of their communist governments. This grassroots movement, which spread throughout Eastern Europe, and which was primarily peaceful, led to the end of the Soviet empire and to the collapse of communism in Europe. Western political scientists were amazed that such an occurrence could take place. Few, if any, had imagined that the end of a powerful total-itarian state could come from the nonviolent actions of average citizens.

A spontaneous grassroots movement also occurred in Argentina in the early 1980s when a group of mothers met daily in one of the main squares in the nation's capital to protest the disappearance of their children (thousands of individuals who were abducted by the military govern-ment in its war against subversion and suspected subversion). The silent, nonviolent protest by the mothers helped undermine the internal and external support for the government.

Advocates of the civil society approach to development say it is easy to show examples of failures by the market and by the state to make peo-ple's lives better. It is even easy to show examples where they have made the people's lives worse. People have responded to the failures of the market and the state by undertaking self-help activities. Such individu-als want to participate in controlling their lives and do not want to let the market or the state be the main determinants of how they should live. They believe that strong reliance on the market or the state can leave the individual stunted.

The advocates of civil society also point to flourishing voluntary efforts in many countries as evidence of the importance of their approach. Although it is impossible to know exactly how many such groups exist today, the Worldwatch Institute in the United States has found much evidence of civil society in the world in the late 1980s. Here is a partial list of what they found:

- In India tens of thousands of groups, many following the self-help tradition established by Mahatma Gandhi, were involved in promoting social welfare, developing appropriate technology, and planting trees.
- In Indonesia, 600 independent groups worked on environmental protection.
- In Sri Lanka, citizens formed a village awakening movement in which one third of the nation's villages participated: 3 million people were involved in work parties, education and health projects, and in coop-erative crafts.
- In Kenya 16,000 women's groups with 600,000 members were regis-tered in the mid-1980s, many starting as savings clubs.

17

- In Brazil, 100,000 Christian Base Communities existed, their membership totaling 3 million.
- The women's self-help movement in the shantytowns surrounding the capital of Lima, Peru, operated 1,500 community kitchens.
- In the United States in the late 1980s an estimated 25 million people were involved in local actions to protect the environment. [14]

Finally, the advocates of the civil society approach point to the spread of democracy around the world. In the 1980s many developing nations adopted a democratic form of government and with the collapse of the Soviet empire, many former communist countries became democratic. By 1992 more countries had adopted democracy than ever before in the world's history: the Freedom House judged that nearly 70 percent of the world's people lived in free or partly free nations at that time. And it is in democracies that voluntary organizations flourish.

Critics of the civil society approach point out that while small may be beautiful, it can also be insignificant. Even the admirable Grameen Bank of Bangladesh provides only about 0.1 percent of the credit in the country. [15] And the conclusion of a UN organization sympathetic to the efforts of self-help groups is that while nongovernmental organizations have helped transform the lives of millions of people throughout the world, "What seems clear is that even people helped by successful projects still remain poor." [16]

Efforts at the grassroots level directed toward community-managed economic development often fail. The worker cooperative is often the instrument used, but a majority of these survive only a few years. [17] The members of cooperatives where workers come together to purchase and operate a business, are usually inexperienced in management. They are plagued by outside economic forces, such as high inflation and uncertain markets, which they are unable to deal with.

Critics also point out that oppressive political and economic powers can block the efforts of community groups. One well-publicized example was the assassination of Chico Mendes, the leader of a group of rubber tree harvesters in the Brazilian Amazon region. The large landowners in this region and in other Latin American countries have, with the support of local governments, traditionally used force against peasant and workers' organizations.

Finally, critics of the civil society approach point to the spread of antidemocratic forces in the world. With the spreading of democracy, the end of the Cold War, and the collapse of communism in Europe, ethnic and regional hatreds have surfaced in many countries, hatreds that had been supressed by the former authoritarian and totalitarian governments. Yugoslavia entered into a cruel civil war and the world saw "ethnic cleansing" reemerge, an idea that it had incorrectly believed had been confined in Europe within Nazi Germany. Bitter ethnic hostilities also arose in Africa in the mid-1990s, with thousands slaughtered in horrifying civil wars. Sometimes incited by a few people for political rea-

sons, group hatred toward "others", toward those outside one
unfortunately has become fairly common in the post-Cold Wa
True civil society, where people have respect and tolerance f
outside their immediate group, does not exist in a number of c
today. [18]

Conclusions

The market approach to development
places emphasis on the seemingly strong
motivation individuals have to acquire
more material goods and services. When
people are freed from external restraints,
the market allows them to use their initiatives to better their lives. The release of
creative energy that comes with the market
approach is impressive. At the end of the
twentieth century most countries throughout the world were following it, at least to
some degree, as the Western capitalist
countries became the models to imitate.

With the collapse of communism and
the breakup of the former Soviet Union,
the state approach to development received
a serious blow. The reliance on the state to
create wealth was discredited. Yet in no
country of the world is the state without
some significant functions relating to the
economy. Within the capitalist world there
is a debate among nations regarding how
much involvement government should
have in directing and guiding the economy.
Japanese capitalism relies on much more
government involvement than does the
capitalism of the United States. And both
the United States and Japan are watching
each other closely and experimenting with
some features that the other utilizes. Even
after the seemingly total victory of the
market approach over the state approach
in the 1990s, the state approach is not
dead; what is dead is the total or near total
reliance on it as the best way to create
wealth.

The market approach relies on the
materialistic self-interest of people to create
wealth, while the state approach presents

the government as
creation of wealth and the reduction of
poverty. The civil society approach presents a new participant in development
and new motivations. By focusing on the
benefits that occur when people exercise
local initiative and function as a community,
a new force is recognized as having a role
in development. Civil society recognizes
that social development – how people
interact – is as important as economic and
political development for the health of a
community. The millions of voluntary
efforts that can take place every day in a
civil society, most of them not motivated
by economic gain or the result of government intervention, can make the difference between a highly civilized community and one that is not.

The challenge societies face is to
achieve the right balance among the three
approaches. It appears that the sole
reliance on only one, or even two, will not
produce a society where the people are living up to their full potential. The United
States can be considered a highly developed society from the perspective of the
market and the state – in other words,
from an economic and political perspective – but underdeveloped as a civil society. The breakdown of the traditional
family is just one indicator of a serious
deficiency in its social development. In
1993 nearly 30 percent of all children in
the United States lived with a single parent
who had never married, an increase of 70
percent from ten years before. [19] And in all
industrial countries the number of
divorces is now one-third the number of
marriages contracted. (Recent studies
indicate that children of divorced parents

19

much more than is widely rec-
) The current state of the family
ety's most basic and important com-
ent – in the United States, along with
many other indicators of ill social health
(such as high crime rates, drug use, and
domestic violence), indicate that not
enough attention is being paid to civil society.
Social critics such as Alan Wolfe, dean of
the New School for Social Research,
believe that the market has become such a
dominant force in the United States that it
is actually weakening the civil society.[21]

Let's return to the central question
addressed in this chapter: What makes
some nations rich and some poor? The
United States and Western Europe do not
appear to offer many relevant lessons for
the poorest nations. Industrialization in
the West took place under conditions
vastly different from those now experienced by many poor nations. The Western
nations were generally rich in natural
resources in relation to their needs, or if
deficient, were able to get them from their
colonies. Most developing nations today
are not rich in the natural resources
needed for industrialization (such as
sources of energy), nor are they permitted
to seize colonies. In addition, the West,
while experiencing an expansion of its
population during its early development,
did not come close to experiencing the
vast population growth that is common
today in many of the poorest nations. And
citizens of the United States especially
need to remember that the unique features
their nation possesses – fertile land, rich
natural resources, an abundance of fresh
water in its numerous lakes and rivers, and
a temperate climate – make their country
atypical in the world. North Americans
have used these gifts to create an unprecedented amount of material wealth—but
with real costs, many of which will be
examined in this book. The policies advocated by the market approach generally

worked well in the United States, where
individual initiative was encouraged by a
supportive government with limited powers.

Since much of the rest of the world does
not share the assets Western Europe and
the United States possess, what should
they do to raise the living standards of the
poorest? Some might advocate revolution
to break the repressive bonds common in a
number of societies, but the suffering that
revolutions cause and the uncertainty of
their accomplishments make it difficult to
advocate this course. But to counsel moderation and slow reforms also is difficult
since, as historian Barrington Moore Jr
reminds us, it ignores the suffering of those
who have not revolted. Moore believes that
"the costs of moderation have been at least
as atrocious as those of revolution, perhaps a great deal more." (Moore also
makes some harsh criticisms of revolutions: "one of the most revolting features of
revolutionary dictatorships has been their
use of terror against little people who were
as much victims of the old order as were
the revolutionaries themselves, often more
so.")[22]

The poor nations should study carefully
the experiences of Japan, South Korea,
Hong Kong, Taiwan, and Singapore. These
lands share with many of the poorest
nations limited natural resources and large
populations. A high regard for education
and hard work in these Asian societies as
well as stable governments have undoubtedly aided them. Government played a
more active role in promoting economic
activity in these countries than called for
by the market approach favored by the
United States. Current research suggests
that a large factor in their economic success was their decision to emphasize uni-
versal education and efforts such as health
care to improve their human capital. The
gap between the rich and the poor was
reduced in these countries, not by taking
excessive wealth from the rich, which

might have stifled entrepeneurship and investments, but by giving the poor the means and the incentive to earn more income. Reducing the income inequality in their countries probably also contributed to political stability, which enabled their governments to stick with sound economic policies.[23] In contrast with Latin America, where the ownership of land is highly unequal (about 1 percent of the landowners there own about 40 percent of the arable land), land reform in Japan, South Korea, and Taiwan reduced the inequality of land distribution in those countries and made more land available to small farmers.[24]

Where does this leave us? It should leave us, I believe, with a sense of humility, if we are the rich, as we recognize how important factors outside of our personal control probably were in ensuring our richness. If we are the poor, it can leave us with an understanding of some of the causes of poverty and some suggestions of how nations might improve the lot of their poorest. And for both the rich and the poor, there should be an awareness that this is the first time in human history that there is a global concern with issues of development – why some are rich and many are poor. That we have not yet learned how to reduce the vast inequalities of wealth in the world should not be surprising. We may never learn how the South can catch up to the North. It seems likely now that within our lifetimes the gap between the rich and poor in the world will increase instead of decrease. But it is clear we are learning, through trial and error, how to improve the lot of the poorest. Whether the poor and rich nations will have the political will – and ability – to do what is necessary to help the world's poor, is not known.

The economies of many of the poor and rich nations are closely linked today. The rich industrial nations of the North need the resources and markets of the poor nations of the South, and the poor nations need to sell their products to the North. And there is another way they are linked. Private banks and governments of the industrial countries and international agencies have lent huge amounts of money to the less developed nations – estimated to have been $1.4 trillion in 1992 – and the poorer nations are having great difficulty repaying these loans.[25] The interest payments of some Latin American nations are now larger than the total funds they earn from their exports. If one or more of the larger borrowers default on these loans, as some have already come close to doing, it could set off an international banking crisis. Poor and rich nations today are economically tied together and, as the following chapters on population, food, energy, the environment, and technology will show, they are also increasingly affected by a number of common global concerns.

NOTES

1 Lynn H. Miller, *Global Order: Values and Power in International Politics* (Boulder, Colo.: Westview Press, 1985), p. 129.

2 United Nations Development Programme, *Human Development Report 1993* (New York: Oxford University Press, 1993), p. 11.

3 The World Bank, *World Development Report 1992 – Development and the Environment* (New York: Oxford University Press, 1992), p. 29. "Consumption" includes food, clothing, rent, fuel, medical care, education, transportation, and durable and nondurable household goods. Caution should be exercised when using consumption statistics since, as the World Bank states, "Estimating the structure of consumption is one of the weakest aspects of national accounting in low- and middle-income economies" (p. 291).

4 Alan B. Durning, "Ending Poverty." In *State of the World, 1990* (New York: W. W. Norton, 1990), p. 137.

5 World Bank, *World Development Report 1990 – Poverty* (New York: Oxford University Press, 1990), p. 2.

6 Current thinking among Western economists about economic development is considerably more complex than the simplified view of the market approach presented below.

7 I am indebted to Alan Wolfe for this classification of the three main views of development. He presented his ideas in a paper titled "Three Paths to Development: Market, State, and Civil Society," which was prepared for the International Meeting of Nongovernmental organizations (NGOs) and UN System Agencies held in 1991 in Rio de Janeiro. Some of his views on this subject are contained in his book *Whose Keeper? Social Science and Moral Obligation* (Berkeley: University of California Press, 1989).

8 Irma Adelman and Cynthia Taft Morris, *Economic Growth and Social Equity in Developing Countries* (Stanford: Stanford University Press, 1973), p. 189. Censuses in Brazil have revealed that the percentage of national income going to the top 10 percent of the population was 40 percent in 1960, 47 percent in 1970, and 51 percent in 1980. During the same period, the poorest 50 percent of the population received 17 percent of the national income in 1960, 15 percent in 1970, and only 13 percent in 1980. Thomas E. Skidmore and Peter H. Smith, *Modern Latin America*, 2nd edn (New York: Oxford University Press, 1989), p. 180.

9 For a fuller discussion of the dependency theory, see Bruce Russett and Harvey Starr, *World Politics: The Menu for Choice*, 2nd edn (New York: W. H. Freeman, 1985), ch. 16; and E. Wayne Nafziger, *The Economics of Developing Countries,* 2nd edn (Englewood Cliffs, NJ.: Prentice-Hall, 1990), pp. 11–14.

10 Russett and Starr, *World Politics,* p. 450.

11 Frederic S. Pearson and J. Martin Rochester, *International Relations: The Global Condition in the Late Twentieth Century,* 3rd edn (New York: Random House, 1992), p. 468.

12 David Maybury-Lewis, *Millennium: Tribal Wisdom and the Modern World* (New York: Viking, 1992), p. 265.

13 United Nations Development Programme, *World Development Report 1993*, p. 95. Further information about the Grameen Bank is contained in Alan B. Durning, "Mobilizing at the Grassroots." In *State of the World*, 1989 (New York: W. W. Norton, 1989), pp. 164–5.

14 Alan Durning, "Mobilizing at the Grassroots," pp. 157–8.

15 *Human Development Report 1993*, p. 95.

16 Ibid., p. 94.

17 Alan Durning, "Mobilizing at the Grassroots," p. 163.

18 An interesting analysis of the causes of the many ethnic conflicts taking place in the world in the mid-1990s is contained in Daniel Goleman, 'Amid Ethnic Wars, Psychiatrists Seek Roots of Conflict," *New York Times,* late edn (August 2, 1994) pp. C1 and C13.

19 Steven A. Holmes, 'Birthrate for Unwed Women Up 70% Since '83, Study Says," *New York Times*, national edn (July 20, 1994), p. A1.

20 Barbara Dafoe Whitehead, "Dan Quayle Was Right," *The Atlantic* (April 1993), pp. 47–84.

21 Alan Wolfe, *Whose Keeper?*

22 Barrington Moore Jr, *Social Origins of Dictatorship and Democracy: Lord and Peasant in the Making of the Modern World,* (Boston: Beacon Press, 1966), pp. 505–7.

23 Sylvia Nasar, "Economics of Equality: A New View," *New York Times,* national edn (January 8, 1994), p. 17.

24 Forces opposing land reform are strong in most countries. Land reform in Japan came only after its defeat in World War II.

25 For a fuller discussion of the debt issue, see Henry L. Bretton, *International Relations in the Nuclear Age: One World Difficult to Manage* (Albany: State University of New York Press, 1986), pp. 232–7, and 290–6; Jim MacNeill, "Strategies for Sustainable Economic Development," *Scientific American,* 261 (September 1989), pp. 156 and 164; and Pearson and Rochester, *International Relations,* p. 493. As Pearson and Rochester state, the huge international debt of the poorer countries is "traceable to their own mismanagement, to high interest rates and generally worsening global economic conditions, and to questionable bank lending practices." In order to earn enough to repay the debt, many poorer countries are promoting environmentally destructive export industries and cutting social welfare expenditures. Industrial countries' tariffs and subsidies for their farmers make it difficult for the poorer nations to sell their products and earn hard currencies.

Further Readings

Broad, Robin, John Cavanagh, and Walden Bello, "Development: the Market Is Not Enough," *Foreign Policy,* 81 (Winter 1990–1), pp. 144–62. After presenting the failures of the market approach, the authors argue that a new type of development is starting to take place in the 1990s, which is emphasizing ecological sustainability, equity, and citizen participation in addition to raising living standards.

Dewey, Clive (ed.), *The State and the Market* (Riverdale, Md: Riverdale Company, 1987). A highly readable collection of essays concerning the role that the world market has played in the economic and social development of the Third World.

Durning, Alan B., "Ending Poverty," In *State of the World,* 1990 (New York: W. W. Norton, 1990), pp. 135–53. A senior researcher at the Worldwatch Institute explores poverty around the world and the ways it can be combated.

Harrison, Lawrence E., *Who Prospers? How Cultural Values Shape Economic and Political Success* (New York: Basic Books, 1992). Examining the effects of culture on development, Harrison maintains that those cultures that place a high value on work, frugality, ties to the whole community, planning for the future, and education are the ones that prosper.

Mittelman, James H., *Out from Underdevelopment* (New York: St Martin's Press, 1988). This look at Third World development examines market and state approaches to economic growth and suggests that an alternative approach combining elements of both views is the best path for underdeveloped nations.

Rich, Bruce, *Mortgaging the Earth: The World Bank, Environmental Impoverishment, and the Crisis of Development* (Boston: Beacon Press, 1993). Rich maintains that because the World Bank believes that central planners can devise suitable development approaches that can be applied throughout the world, the Bank has fostered projects that have caused major environmental destruction and dislocated millions of people.

de Soto, Hernando, *The Other Path: The Invisible Revolution in the Third World* (New York: Harper and Row, 1989). De Soto focuses on the informal part of the economy in Peru, those illegal street vendors, private bus owners, and slum dwellers who have built houses on illegally occupied land. His provocative thesis is that these people are the real and only capitalists in the country and are the most productive part of the economy.

Stone, Roger D., *The Nature of Development: A Report from the Rural Tropics on the Quest for Sustainable Economic Growth* (New York: Alfred A. Knopf, 1992). The author visited rural development projects in Latin America, Asia, and Africa to attempt to discover why some succeeded and some failed in their attempts to develop small-scale but economically sustainable projects.

United Nations Development Programme, *Human Development Report 1993* (New York: Oxford University Press, 1993). This uncharacteristically interesting report rates nations according to their level of human development and focuses on the participation of people throughout the world in efforts to improve their living conditions.

Vogel, Ezra F., *The Four Little Dragons: The Spread of Industrialization in East Asia* (Cambridge: Harvard University Press, 1991). This examination of four of the most economically successful places in East Asia – South Korea, Taiwan, Singapore, and Hong Kong – seeks to explain how industry and government interact in each one.

World Bank, *World Development Report 1990 – Poverty* (New York: Oxford University Press, 1990). This particular report focuses on poverty. It presents the policies, politics and programs affecting poverty and presents a strategy to reduce it. The book's many statistics are presented in colorful charts.

World Resources Institute, *World Resources 1994–95* (New York: Oxford University Press, 1994). A valuable, easy to read reference source detailing the resources available to rich and poor countries, their condition, and the trends in their use. The book provides an opportunity for contrast and comparison of developed and developing nations.

2

Population

The Changing Population of the World ● ● ● ● ● ● ● ● ●

The population of the world is growing. No one will be startled by that sentence, but what is startling is the rate of growth, and the fact that the present growth of population is unprecedented in human history. The best historical evidence we have today indicates that there were about 5 million people in the world about 8000 BC. By AD 1 there were about 200 million, and by 1650 the population had grown to about 500 million. The world reached its first billion people about 1800; the second billion came about 1930. The third billion was reached about 1960, the fourth about 1975, and the fifth about 1987. The sixth is expected to come about 1998. These figures indicate how rapidly the population is increasing. Table 2.1 shows how long it took the world to add each billion of its total population.

There is another way to look at population growth, one that helps us understand the uniqueness of our situation and its staggering possibilities for harm to life on this planet. Because most people born can have children of their own, the human population can – until certain limits are reached – grow exponentially: 1 to 2; 2 to 4; 4 to 8; 8 to 16; 16 to 32; 32 to 64; 64 to 128; etc. When something grows exponentially, there is hardly discernible growth in the early stages and then the numbers shoot up. The French have a riddle that they use to help teach the nature

Table 2.1 Time taken to add each billion to the world population, 1800–1998

Date	Estimated world population(billions)	Years to add 1 billion people
1800	1	2,000,000
1930	2	130
1960	3	30
1975	4	15
1987	5	12
1998 (projected)	6	11

of exponential growth to children. It goes like this: if you have a pond with one lily in it that doubles its size every day, and which will completely cover the pond in 30 days, on what day will the lily cover half the pond? The answer is the twenty-ninth day. What this riddle tells you is that if you wait until the lily covers half the pond before cutting it back, you will have only one day to do this – the twenty-ninth day–because it will cover the whole pond the next day.

If you plot on a graph anything that has an exponential growth, you get a J-curve. For a long time there is not much growth but when the bend of the curve in the "J" is reached, the growth becomes dramatic. Figure 2.1 shows what the earth's population growth curve looks like.

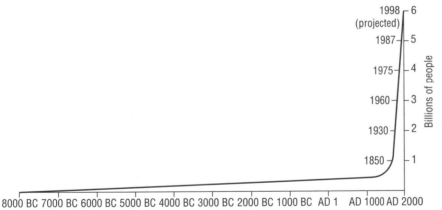

Figure 2.1 Population growth from 8000 BC to 1998

The growth of the earth's population has been compared to a long fuse on a bomb: once the fuse is lit, it sputters along for a long while and then suddenly the bomb explodes. This is what is meant by the phrases "population explosion" and "population bomb." The analogy is not a bad one. The world's population has passed the bend of the J-curve and is now rapidly expanding. The United Nations expects the world's population to

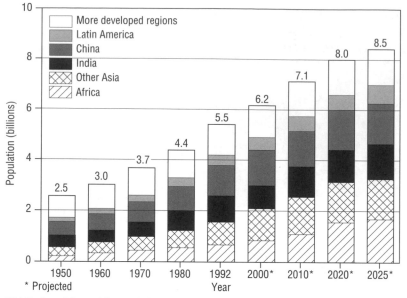

Figure 2.2 Distribution of the world's population, 1950–2025.

Source: UN Department of Economic and Social Information and Policy Analysis, *Current Demographic Situation and Future Change: Selected Pages from World Population Prospects, the 1992 Revision* (New York: United Nations, 1993) p.14.

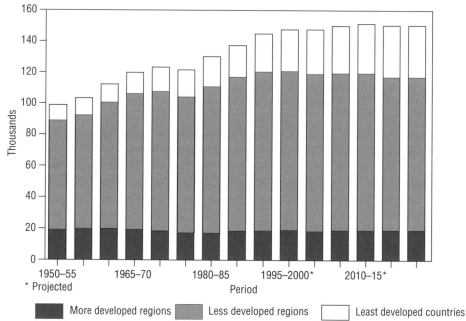

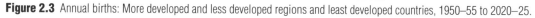

Figure 2.3 Annual births: More developed and less developed regions and least developed countries, 1950–55 to 2020–25.

Source: UN Department of Economic and Social Information and Policy Analysis, *Current Demographic Situation and Future Change: Selected Pages from World Population Prospects, the 1992 Revision* (New York: United Nations, 1993), p.21.

reach over 6 billion by the year 2000, shooting up from the estimated 4 billion in 1975, an increase of more than 50 percent in just 25 years. Figure 2.2 gives the changing distribution of the world's population and projections to the year 2025.

The figure also shows that the largest growth in the future will be in the poorest countries of the world. Another way to show this is given in figure 2.3 which provides data on annual births in more developed and less developed regions.

Between 1950 and 1990, about 85 percent of the growth of the world's population occurred in the less developed regions and about 15 percent in the more developed regions. Over the next 35 years, about 95 percent of population growth will occur in the less developed countries. An ever larger percentage of the world's population will be nonwhite and poor. (A number of African countries are growing somewhat more slowly than expected because of the AIDS epidemic, but the UN's expectation in the early 1990s was that the epidemic would have only a modest effect on Africa's population size.[1]) In 1950 about two-thirds of the world's people lived in the less developed countries. In the early 1990s this figure had increased to about 75 percent and the United Nations projects that by 2025 about 85 percent of the earth's population may be residing in the poorer nations.

High growth rates will take place in the less developed countries because a large percentage of their population consists of children under the age of 15 who will be growing older and having children themselves. If we plot the number of people in a country according to their ages, we can see clearly the difference between rapidly growing populations, which most less developed nations have, and relatively stable or slowly growing populations, which the more developed nations have. Figure 2.4 shows the difference between the populations of more developed and less developed nations of the world in 1975 and 2000. The age structure of the more developed countries is generally column-shaped, while the age structure of the less developed countries is usually pyramid shaped. The United States, as a fairly typical more developed country, has a population distribution similar to the top figure, while Mexico, a fairly typical less developed country, has a population distribution similar to the bottom.

Another major change occurring in the world's population is the movement of people from rural to urban areas. Although this is happening throughout the world, the trend is especially dramatic in the Third World, where people are fleeing rural areas to escape the extreme poverty that is common in those areas, and because the cities seem to offer a more stimulating life. Mostly it is the young people who go to the cities, hoping to find work and better living conditions. But all too often jobs are not available in the cities either. These rural migrants usually settle in slums on the edges of the big cities. It is estimated that from 30 to 60 percent of the urban populations of Third World countries live in such shantytowns (called "uncontrolled settlements" in government

29

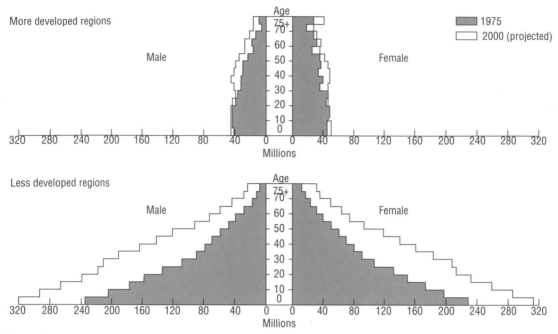

More developed regions

1975
2000 (projected)

Age
Male   Female

Less developed regions

Age
Male   Female

Figure 2.4 Age and sex distribution of population of more developed and less developed regions.

Source: Council on Environmental Quality and the Department of State, *The Global 2000 Report to the President* , vol. 1 (Washington: *Government Printing Office, 1980),p.11.*

Rural migrants often settle in urban slums in developing nations. *(United Nations)*

reports).[2] At current rates the populations in these informal settlements will double every 10 to 15 years. It is hard to imagine a city like Calcutta getting any bigger. In 1950 it had a population of about 4 million, and thousands of these people lived permanently on the streets; in 1990 it had a population of about 11 million and an estimated 400,000 lived on the streets.[3] If the present rate of growth continues, it will have a population of about 16 million by 2010. Table 2.2 gives the world's ten largest cities in 1950 and 1990 and the projected ten largest for the year 2000[4]. Note that seven of the ten largest cities in the year 2000 are expected to be in the less developed countries – Tokyo, New York, and Los Angeles are the exceptions – whereas in 1950 only three of the ten (Shanghai, Buenos Aires, and Calcutta) were in the poorer countries.

Table 2.2 Ten largest cities in the world, 1950, 1990, and 2000

	Population in 1950 (millions)		Population in 1990 (millions)		Projected population in 2000 (millions)	
1	New York City	12	Tokyo	25	Tokyo	28
2	London	9	Sao Paulo	18	Sao Paulo	23
3	Tokyo	7	New York City	16	Bombay	18
4	Paris	5	Mexico City	15	Shanghai	17
5	Moscow	5	Shanghai	13	New York City	17
6	Shanghai	5	Bombay	12	Mexico City	16
7	Essen, Germany	5	Los Angeles	12	Beijing	14
8	Buenos Aires	5	Buenos Aires	11	Lagos	14
9	Chicago	5	Seoul	11	Jakarta	13
10	Calcutta	4	Rio de Janeiro	11	Los Angeles	13

Source: United Nations Department of Economic and Social Information and Policy Analysis, *World Urbanization Prospects: The 1992 Revision,* (New York: United Nations, 1993), pp. 13–14.

Note also the increased size of the cities. Cities with over 5 million people are sometimes called "megacities." In 1950 Buenos Aires and Shanghai were the only cities in the developing world with at least 5 million people. By 2000 there are expected to be as many as 35 megacities in the Third World. Many of these cities will have vast areas of substandard housing and serious urban pollution, and many of their residents will live without sanitation facilities, safe drinking water, or adequate health care facilities.

The world's population is becoming increasingly urban. Although countries differ on their definitions of "urban" (the United States defines urban as places with populations of 2,500 or more, Japan uses 50,000, and Iceland 200), in 1994 about 75 percent of the people in the more developed nations lived in urban areas whereas about 35 percent of the people in the less developed countries were urban. In 1950 only about 15 percent of the population in the poorer nations was urban. The United Nations expects that by 2015 about half of the population in the poorer nations will live in urban areas. By 2005, for the first time in history, more people in the world will live in urban than in rural areas.

Growing cities in less developed nations often have a mixture of modern and substandard housing. *(United Nations)*

Causes of the Population Explosion

Although it is easy to illustrate that the human population is growing exponentially, it is not so easy to explain why we are in a situation at present of rapidly expanding population. Exponential growth is only one of many factors that determine population size. Other factors influence how much time will pass before the doublings – which one finds in exponential growth – take place. Still other factors influence how long the exponential growth will continue and how it might be stopped. We will consider these last two matters later in the chapter, but we will first look at some of the factors that drastically reduced the amount of time it took for the world's population to double in size.

The Agricultural Revolution, which began about 8000 BC, was the first major event that gave population growth a boost. When humans learned how to domesticate plants and animals for food, they greatly increased their food supply. For the next 10,000 years, until the Industrial Revolution, there was a gradually accelerating rate of population growth, but overall the rate of growth was still low because of high death rates, caused mainly by diseases and malnutrition. As the Industrial Revolution picked up momentum in the eighteenth and nineteenth centuries, population growth was given another boost as advances in industry, agriculture, and transportation improved the living conditions of the average person. Population was growing exponentially, but the periods between the doublings were still long because of continued high death rates. This situation changed drastically after 1945. Lester Brown of the Worldwatch Institute explains why that happened:

> The burst of scientific innovation and economic activity that began during the forties substantially enhanced the earth's food-producing capacity and led to dramatic improvements in disease control. The resulting marked reduction in death rates created an unprecedented imbalance between births and deaths and an explosive rate of population growth. Thus, while world population increased at 2 to 5 percent *per century* during the first 15 centuries of the Christian era, the rate in some countries today is between 3 and 4 percent *per year*, very close to the biological maximum.[5]

It was primarily a drastic reduction in the death rate around the world after World War II that gave the most recent boost to population growth. The spreading of public health measures, including the use of vaccines, to Third World countries enabled these countries to control diseases such as smallpox, tuberculosis, yellow fever, and cholera. Children and young adults are especially vulnerable to the infectious diseases; thus the conquering of these diseases allowed more children to live and bear children themselves.

While death rates around the world were dropping rapidly, birth rates remained generally high. Birth rates have been high throughout human history. If this had not been true, you and I might not be here today since high birth rates were needed to replenish the many people who died at birth or at an early age. (If you walk through a very old cemetery in the United States or especially in Europe, you can see evidence of this fact for yourself as you pass the family plots with markers of the many children who died in infancy and in adolescence.) Birth rates remained high right up until the late 1960s, when a lowering of the rate world-wide was seen, which was probably the beginning of a gradual lowering of the birth rate around the world.

The birth rate has dropped significantly in the developed nations but remains high in most Third World countries. There are a number of reasons for this. First, many poor people want to have many children. If many of them die in infancy, as they still do in countries such as India, Pakistan, Bangladesh, and in tropical Africa, many births are needed so

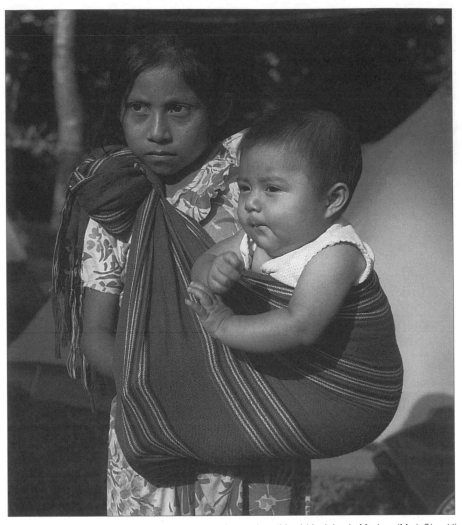

Children take care of children in many poorer countries, such as this girl is doing in Mexico. *(Mark Olencki)*

that a few children survive. If the poor families are peasants, as many of them are, sons are needed to work in the fields and to do chores. And since there are rarely old age pension plans in Third World countries, at least one son (and preferably two) is needed to ensure that the parents have someone to take care of them when they are old and can no longer work. These needs are reflected in the commonly heard greeting to a woman in rural India: "May you have many sons."

Poor families also want children to provide extra income for the family. Before child labor laws severely restricted the use of children in factories in the United States and Europe, it was common for children to take paying jobs to help the family gain income. A study of an Indian village found that the poorest families had many children since they felt that increased numbers meant more security and a better chance for prosperity.[6]

Other reasons for continued high birth rates in poor countries are tradition and religion. The unusually high birth rates in large parts of Africa today are primarily caused, according to one study, by family patterns and religious beliefs that developed over thousands of years in response to conditions in the region.[7] Tradition is very important in most rural societies, and traditionally families have been large in rural settings and among poor people. One does not break with such a tradition easily. Also, religion is a powerful force in rural societies and some religions advocate large families. The influence of Islamic fundamentalism is strong in some Islamic states and it is a major force discouraging the use of contraceptives by women. The Catholic religion is a powerful force in rural Latin America, especially with the women, and many – although fewer than in earlier times – obey the Catholic Church's prohibition against the use of contraceptives. In some Latin American countries men have commonly regarded a large number of children as proof of their masculinity.

The unavailability of birth control devices is also a reason for the high birth rates in developing nations. It has been estimated that about 120 million married women of reproductive age worldwide are not using contraceptives even though they do not want more children. It is believed that these women have an unmet need, or demand, for family planning services. That number represents one woman out of every five in the developing world, outside of China.[8]

How Population Growth Affects Development ● ● ● ● ● ● ● ● ● ●

How does population growth affect development? Although there is no easy answer to the question of what is "too large" or "too small" a population for a country – a question we will return to in the final section of this chapter – we can identify some obvious negative features of a rapidly growing population, a situation that would apply to many Third World countries today.

Too Rapid

Let's look again at the age distribution of the population in less developed regions shown in figure 2.4. It is striking that a large percentage of people are below the age of 15. This means that a large portion of the population in these countries is mainly nonproductive. Although children do produce some goods and services, as mentioned above, they consume more than they produce. Food, education, and health care must be provided to them before they become old enough to become productive themselves. Obviously, if a nation has a large portion of its population in the under-15 age group, its economy will be faced with a huge burden.

A rapidly growing population also puts a great strain on the resources of the country. If the population is too large or the growth too rapid, people's use of the country's resources to stay alive can actually prevent the biological natural resources from renewing themselves. This can lead to the land becoming less fertile, and the forests being destroyed. An example of this is the making of patties out of cow droppings and straw by women in India and Pakistan. These patties are allowed to dry in the sun and are then used for fuel. In fact, dung patties are the only fuel many peasants have for cooking their food. But the use of animal droppings for fuel prevents essential nutrients from returning to the soil, thus reducing the soil's ability to support vegetation.

A large population of young people also means that there will be a terrific demand for jobs when these children grow old enough to join the labor force – jobs that are unlikely to exist. The ranks of the unemployed and underemployed will grow in many Third World nations, and this can easily lead to political and social unrest. As we saw earlier in this chapter, people from the rapidly growing rural areas of the Third World are heading for the cities hoping to find work. What they find though is a scarcity of jobs, undoubtedly a contributing factor in the high rates of urban crime.

Urban crime in the third world: a personal experience

An experience in Liberia helped me to understand that urban areas in less developed nations are often less safe than rural areas. I lived at different times in both Monrovia, the capital city of that country and in a small village in a rural area. Once while I was in Monrovia, a thief entered my bedroom and stole my wallet and watch from under my pillow, which was under my sleeping head at the time. Such an event was unheard of in the rural areas, but was not that uncommon in the city. After the theft happened, I was happy to return to my "primitive" village where I felt much safer.

Rapid population growth also has a harmful effect on the health of children and women. Malnutrition in infancy can lead to brain damage, and child-bearing frequently wears women down. This is what happens to many Third World women:

> After two decades of uninterrupted pregnancies and lactation women in their mid-thirties are haggard and emaciated, and appear to be in their fifties. As researchers Erik Eckholm and Kathleen Newland point out, such women are "Undernourished, often anemic, and generally weakened by the biological burdens of excessive reproduction," they "become increasingly vulnerable to death during childbirth or to simple infectious diseases at any time," and " their babies swell the infant mortality statistics."[9]

A rapidly growing population also puts a tremendous strain on the ability of a nation to provide housing for its people. The poor condition of much of the housing in the Third World is something that makes a lasting impression on foreign visitors to these countries – that is, if they venture beyond the Hilton Hotels where they often stay. Overcrowding also is produced by an excessive and rapidly growing population, and that leads to a scarcity of privacy and to limited individual rights.

Not surprisingly, a study sponsored by the US National Academy of Sciences to explore the relationship between population growth and economic development concluded in the mid-1980s that slower population growth would aid economic development in most of the less developed nations.[10]

Too Slow

A country's population growth rate can also be too slow to support a high level of economic growth. Partly because of low birth rates, a number of European countries had to import unskilled workers during the 1950s and 1960s from Turkey, southern Italy, and other relatively poor areas of Europe and North Africa. Within the world of business, there is concern if the population stops expanding since a growing population is seen as representing more consumers of products. But a number of the industrial countries have shown in the post-World War II period that a high level of economic growth can be obtained even when population growth is low.

Japan is a good example to look at. Even though the country has experienced impressive economic growth, the Japanese show some ambivalence regarding its extremely slow population growth. A survey conducted in 1990 by the Japanese government revealed that many people believed that a recent decline in fertility rates was undesirable because it could lead to an aging population and fewer young people entering the labor market. At the same time the survey revealed that many people (nearly one half) believed that the country was overpopulated.[11] This latter view is probably not surprising given the fact that much of the land is mountainous, with only about 15 percent of the land being suitable for cultivation. About one-quarter of the nation's population lives in the Tokyo metropolitan area, with the consequence that the cost of land and housing in that area is extremely high.

An Aging Population and Low Birth Rates

We saw earlier the types of problems that are created when a country has a large portion of its population aged 15 or under. But special problems are also created when the proportion of a population that is over 65 starts to expand. The United States is facing such a problem with its Social Security system, which provides financial support to retired persons. As the percentage of the US population that is over 65 expands because of advances in health care, and the number of new workers is reduced because of low birth rates, the ratio of working-age people to retired people declines and puts a strain on the Social Security system. (It is the payments from the current workers that provide money for the retirement benefits.)

There are also increased governmental health care costs as a population ages. More funds are needed to care for the medical and social needs of the aged since most developed countries believe it is the whole community's responsibility to help families pay for these services. This is a common concern in Europe where by the year 2020 it is expected that in the countries of the European Community 1 in every 4 people will be over 60, whereas at present fewer than 1 in every 5 are of that age. Even of more concern is the expected doubling of the number of people over 75, from the current 6 percent to over 12 percent.[12]

When a country has a low birth rate, and the number of young people entering the labor market is reduced – a situation now common throughout Europe – often the result is conflict over immigration policies. Hostility to foreign workers by extreme nationalists in Germany in the early 1990s led to fatal attacks on some foreigners in the country. Japan is also concerned with having to rely on foreign workers. (And the Japanese are as concerned as the Americans and Europeans are that a shrinking work force will be unable to support the increasing health care and welfare costs of an aging population.)

A number of European nations had such low birth rates in the mid-1990s that their populations had started to decline or would soon do so. Declining populations also became common in Russia and the former Eastern European satellites, no doubt because of the harsh economic conditions these countries were facing as they tried to replace their planned economies with market economies. Declining populations raised fear about the loss of national power, economic growth, and even national identities by some people in these countries. But most population experts believe that if population decline is gradual, its negative social and economic consequences can be handled. Much more difficult to manage, they believe, are situations where the decline is rapid.[13]

The Relationship Between Population Growth and Poverty

The first international conference on population was held in 1974 in Romania under the sponsorship of the United Nations. It was anticipated that this conference would dramatize the need for population control programs in the Third World, but instead a debate took place between rich and poor countries over what was causing poverty: population growth or underdevelopment. The United States and other developed nations argued for the need for birth control measures in the poorer countries, while a number of the poorer countries argued that what was needed was more economic development in the Third World. Some developing countries called for a new international economic order to help the Third World develop. They advocated more foreign aid from the richer countries, and more equitable trade and investment practices. The conference ended with what seemed to be an implicit compromise: that what was needed was both economic development and population control, that an emphasis on only one factor and a disregard of the other would not work to reduce poverty.

In 1984 the United Nations held its second world population conference in Mexico City. The question of the relationship between economic growth and population growth was raised again. The United States, represented by the Reagan administration, argued that economic growth produced by the private enterprise system was the best way to reduce population growth. The United States did not share the sense of urgency that others felt at the conference concerning the need to reduce the world's increasing population. It announced that it was cutting off its aid to organizations that promote the use of abortion as a birth control technique. (Subsequently the United States stopped contributing funds to the United Nations Fund for Population Activities and the International Planned Parenthood Federation, two of the largest and most effective organizations concerned with population control.)[14] The United States stood nearly alone in its rejection of the idea that the world faced a global population crisis as well as in its advocacy of economic growth as the main population control mechanism. The conference endorsed the conclusion reached at the first conference ten years earlier that *both* birth control measures and efforts to reduce poverty were needed to reduce the rapidly expanding population of the Third World.

In 1992 the United Nations conference on the environment and development – the so-called Earth Summit held in Rio de Janeiro, Brazil, which will be discussed in detail in chapter 5 – did not directly address the need for population control measures. The Rio Declaration says only that "states should ... promote appropriate demographic policies," and Agenda 21, the action plan to carry out the broad goals stated in the declaration, does not mention family planning. The weak treatment of the population issue by this conference was the result of North/South conflicts over whether the poor nations or the rich nations were mainly responsible for the destruction of the environment. (When the population issue was raised, attention was focused on the harm to the environment that large numbers of poor people in the South could inflict, whereas the South held that overconsumption by the North caused most of the pollution that was harming the environment.) The failure to directly address the connection between rapid population growth and environmental damage was also a result of opposition by the Vatican to any declarations that could be used to support the use of contraceptives and abortion to control population growth. Opposition was also voiced by some countries with conservative social traditions concerning issues that could raise the subject of the status of women in their countries.

In spite of the failure to strongly address the population issue in its formal statements, the Rio conference, and the multitude of meetings around the world held to prepare for it, did cause increased attention to be placed on population, especially bringing to the forefront the perspectives of women.

The United Nations held its third conference on population – formally called the International Conference on Population and Development – in Cairo, Egypt in 1994. Although the Vatican and conservative Islamic governments made abortion and sexual mores the topic of discussion in the early days of the conference, the conference broke new ground in agreeing that women must be given more control over their lives if population growth is to be controlled. The conference approved a 20-year plan of action whose aim is to stabilize the world's population at about 7.3 billion by 2015. The plan calls for new emphasis to be placed on the education of girls, providing a large range of family planning methods for women, and providing health services and economic opportunities for females. The action plan called for both developing and industrial nations to increase the amount they spend on population-related activities to $17 billion by the year 2000, a significant increase over the $5 billion that is now spent. Whether or not countries will actually follow through with the conference's recommendations and commit these funds for this purpose is unknown. As the secretary general of the conference, Dr Nafis Sadik, stated at the conclusion of the conference, "Without resources the Program of Action will remain an empty promise." [15]

How Development Affects Population Growth ● ● ● ● ●

How does development affect the growth rate of population? There is no easy answer to that question, but population experts strongly suspect that there *is* a relationship, since the West had a fairly rapid decline in its population growth rate after it industrialized. In the nineteenth century Europe began to go through what is called the "demographic transition."

Demographic Transition

The demographic transition has three basic stages. In the first stage, which is often characteristic of preindustrial societies, there are high birth rates and high death rates, which lead to a stable or slowly growing population. In the second stage, which most industrial nations passed through from about the mid-1800s to the mid-1900s, there is a rapid decline in the death rate as modern medicine and sanitation measures are adopted, which is followed, after some delay, by a drop in the birth rate. During this second stage of transition population increases rapidly, since the reduction in the death rate takes place before the lower birth rate. In the final, and third stage, both the death and birth rates are low, and, as in the first stage, there is a stable or slowly growing population.

Figure 2.5 shows birth and death rates for both more developed and less developed nations. The figure shows that the more developed nations are already in the third stage of the demographic transition but that the less developed nations are still in the second stage. It also shows some significant differences between the developed and developing nations' second stages. For the developed nations, the reduction in the death rate was gradual. The birth rate dropped sharply, but only after a delay that caused the population to expand. For the developing countries, the drop in the death rate has been sharper than it had been for the developed nations, and the reduction in the birth rate has lagged more than it had for the developed nations. Both of these facts have caused a much larger increase in the population of the less developed nations than had occurred in the more developed nations. Figure 2.6 shows the population experiences of two countries: Sweden and Sri Lanka (formerly known as Ceylon). Sweden has passed through the demographic transition and has now reached zero population growth. Sri Lanka is experiencing a much more dramatic second stage than Sweden ever experienced and is having a much greater increase in population. Sri Lanka's population is still growing rapidly, and it has not had the opportunity to get rid of its excess population by sending it overseas to the United States, as Sweden did.

Such differences in the experiences of Sweden and Sri Lanka, as well as the experiences of other developed and developing nations, have led many demographers to change the opinion they had in the 1950s that economic development would cause Third World nations to go through the same demographic transition – and thus achieve lower population growth – as the West had. There are obviously important differences between the Western experience and that of the Third World. Probably as important as the fact that death rates have dropped much faster in the Third World than they had for the West is the fact that the industrialization that is taking place in much of the Third World is not providing many jobs and is not benefiting the vast majority of people in those

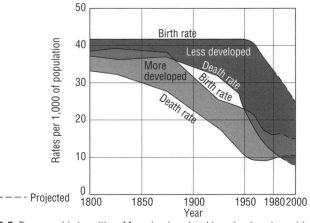

Figure 2.5 Demographic transition: More developed and less developed countries.

Source: Halfdan Mahler, "People," *Scientific American* (September 1980), p. 75.

41

countries. A relatively small, modern sector *is* benefiting from this economic development, and the birth rate of this group is generally declining; but for the vast majority in the rural areas and in the urban slums, the lack of jobs and continued poverty are important factors in their continued high birth rates.

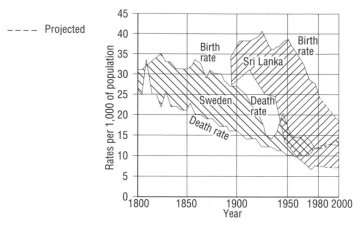

Figure 2.6 Demographic transition: Sweden and Sri Lanka.

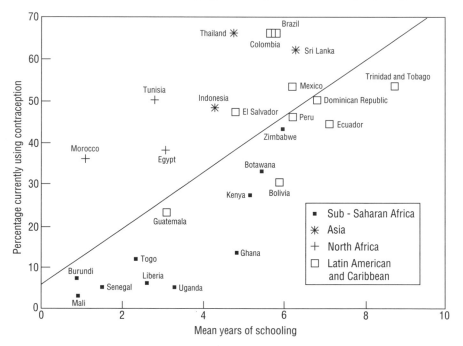

Figure 2.7 Rate of contraceptive use by years of female education.

Source: UN Population Division, *Women's Education and Fertility Behaviour : Recent Evidence from the Demographic and Health Surveys* (New York: United Nations, forthcoming).

42

Factors Lowering Birth Rates

If industrialization as it is occurring in the Third World is not an auto-matic contributor to lower birth rates, what factors do cause birth rates to decline? Certainly, better health care and better nutrition, both of which lower infant mortality and thus raise a family's expectations of how many children will survive, are important factors. (The irony here, of course, is that these advances, at least in the short run, tend to worsen the population problem since more children live to reproduce.) Another factor tending to lower birth rates is the changing role of women. Better educated women are more likely to use some sort of contraception than are those women with little or no education.[16] Figure 2.7 shows the rela-tionship between female education and contraceptive use in 26 less developed countries. Education for women enables them to delay mar-riage, to learn about contraceptives, and to acquire different views of their role in society. And their education allows them to obtain jobs that can often be of more benefit to the family than a larger family.

Actually, however, a little bit of modernization can be a bad thing if birth rates are to be reduced. In large parts of the Third World women with a little exposure to Western ways are giving up breast-feeding of their babies and switching to bottle-feeding. Their fertility goes up when they do this since prolonged breast-feeding naturally delays – some-times for years – a woman's ability to conceive again.

As Western nations industrialized, child labor laws, compulsory edu-cation for children, and old age pension laws reduced the economic incentive for having many children. Children changed from being pro-ducers on the farms and in the early factories and became instead con-sumers and an economic burden on their families. Also, as the West industrialized, it became more urban, and living space became more scarce and more expensive than it was in rural societies. The availability of goods and services increased, which caused families to increase their consumption of these rather than spend their income on raising more children. Traditional religious beliefs, which often support large families, also tended to decline.

There is little debate today that economic development, especially if it benefits the many and not just the few, can lead to lower birth rates. And there is ample evidence that improving the social and economic status of women can lead to lower birth rates, even in areas which remain very poor – such as in the southern state of Kerala in India, where birth rates are significantly lower than in the rest of India. But there is now evidence that birth rates can decrease and are decreasing in poor countries – even in some where there has been little or no eco-nomic growth and where the education and social status of women remains very low – if an effective family planning program exists and modern contraceptives are available.[17] The conclusion of some researchers who have reviewed the results of recent fertility studies conducted in various less developed countries is that "although development and social change create conditions that encourage smaller family size, contraceptives are the best contraceptive."[18] These researchers found that three factors are mainly responsible for the impressive decline in birth rates that have occurred in many Third World countries since the mid-1960s: more influential and more effective family planning programs, new contraceptive technology, and the use of the mass media to educate women and men about birth control.[19] Women throughout the Third World now have four children on average, whereas before the mid-1960s they had six.

Breast-feeding delays a woman's ability to conceive and provides the most healthful food for a baby. *(United Nations).*

Governmental Population Policies ● ● ● ● ● ● ● ● ● ● ● ● ● ● ● ●

Controlling Growth

Many governments today have some policies that try to control the growth of their populations, but this is a very recent trend. Traditionally, governments have sought to increase their populations, either through the encouraging of immigration (as the United States did in its early years) or through tax and other economic assistance to those families with many children. As late as the mid-1970s, many governments had no population control programs. A survey of developing nations taken in conjunction with the 1974 UN population conference found that, out of 110 developing countries, about 30 had population control programs, another 30 had information and social welfare programs, and about 50 had no population limitation programs at all.[20] This UN conference ended with no explicit consensus among the participants that there was a world population problem at all. The delegates at the conference did pass a resolution stating that all families have the right to plan their families and that it is the responsibility of governments to make sure all families have the ability to do so.

Advertisement for contraceptives in Costa Rica. *(George Shiflet)*

The ability to control the number and timing of children a couple has is called family planning. Family planning services provide health care and information on contraceptives. In 1990 about 50 per cent of all couples of childbearing age in the developing world were using some means of birth control. (In the early 1960s only about 20 per cent of women in the less developed countries were practicing family planning.[21]) In the late 1980s the highest use of birth control in the developing world was in East Asia – which includes China, South Korea, Taiwan, and Singapore, among other countries – where an estimated 75 percent were using contraceptives.[22] Contraceptive use in Africa was only about 15 percent in the same period.[23] (The more developed world had a use rate of about 70 percent.) As mentioned earlier in this chapter, in the early 1990s about 120 million women of reproductive age in the poor nations wanted no further children but they were not using contraceptives. They are considered to be potential family planning users if the services were made available to them.

Requests by developing nations for foreign aid to help them control their population growth now, for the first time, exceed the international assistance available for this activity. It is estimated that if all the demand for family planning assistance in the developing world was met, the number of couples using birth control devices would increase from the present 50 percent to over 60 percent.[24] This increased use would result in the average number of children per woman in the developing countries dropping to 3 from the present 4. Providing family planning services to the approximately 120 million women whose potential demand remained unmet at present would cost an estimated $2.4 billion annually. While this seems like a huge amount, relative to other expenditures being made at present it is not. (The cost of one modern submarine in the US is over $2 billion and the US tobacco industry spends about that amount yearly on advertising.) The US government has been the largest single donor of aid for population and family planning activities in the developing world, providing about $4 billion during the 1970s and 1980s. In real dollars (controlling for inflation) US aid in 1990 was actually lower that it was in the late 1970s. In 1992 the $325 million the United States spent on population-related activities represented about 5 percent of the aid it gave for development purposes.

Mexico is a country that has had rather dramatic success with its family planning program. The government began this program only in 1972, when it had one of the highest rates of population growth in the world. In the early 1970s the annual population growth rate was estimated to be 3.2 percent while in 1994 it was estimated to be down to 2.2 percent.[25]

In 1972 Mexico's President Luis Echeverria Alvarez announced a reversal of governmental policy on the population issue. His decision to support a strong effort to control the rapid growth of the Mexican population led the government to use MEXFAM, the local affiliate of the International Planned Parenthood Federation, to set up family planning clinics throughout the country. (By the early 1990s MEXFAM had set up 200 of these clinics.) Besides making contraceptives readily available, the government and MEXFAM mounted a large propaganda campaign using television soap operas, popular songs, billboards, posters on buses and in subway stations, and spot announcements on radio and television. The leaders of the Catholic Church in Mexico did not oppose the government's efforts.

But if the present birth rate is not reduced further, Mexico's population will double in 30 years. To increase the use of contraceptives, the National Population Council is now focusing its efforts on the rural population, adolescents, and men.[26] Men are an especially important target since the rate of contraceptive use by men in Mexico is low. The use of reliable contraceptives by males accounts for only about 15 percent of contraceptive use in less developed countries,[27] and reportedly one half of the women in Mexico using family planning services do not tell their husbands because they fear physical abuse.[28] Partly because of difficult economic conditions in the country, men are receptive to the message that controlling family size makes sense.

Family planning class. *(United Nations)*

A few countries have adopted more forceful measures than family planning to try to reduce their population growth. Japan drastically reduced its population growth by legalizing abortion after World War II. India, which has had disappointing results with its voluntary family planning programs, enacted more forceful measures in the mid-1970s, such as the compulsory sterilization of some government workers with more than two children. Several Indian states passed laws requiring sterilization and/or imprisonment for those couples who bore more than two or three children. A male vasectomy program was also vigorously pursued, with transistor radios and money being given as an incentive to those agreeing to have the sterilization operation. Public resentment against these policies mounted and helped lead to the defeat of Prime Minister Indira Gandhi's government in 1977. Birth control efforts slackened after that event. The Indian government has now returned to voluntary measures to try to limit the growth of its population, which is now increasing by over 1 million a month. India's population is expected to exceed 1 billion by the year 2000.

China, which has about one-quarter of the world's population but only about 7 percent of its arable land, has launched a vigorous program to limit its population growth and has drastically reduced its birth rate. For many years the communist government, under the leadership of Mao Zedong, encouraged the growth of the population, believing that there was strength in numbers. The policy was eventually reversed, and today the government hopes to limit the population to 1.3 billion by the year 2000, although it will still continue to increase after that. (A 1990 national census revealed that there were about 1.1 billion people in the country.) A wide assortment of measures is being used to limit the growth. Contraceptives are widely promoted, sterilization is encouraged, and abortion is readily available. The government, through extensive publicity efforts, began promoting the one-child family as the ideal. Late marriages are strongly encouraged and couples who have only one child receive better jobs and housing, whereas couples with more than two children are taxed more and receive reduced pay. Sociologist Ronald Freedman describes why the Chinese efforts have been successful:

> The massive Chinese national birth-planning program ... has been organized through the network of political and social organization which mobilizes the masses of the population in primary groups at their places of work and residence ... . That system is used to promote priority objectives–such as birth planning – by persistent and repetitive messages, discussions, and both peer and authority pressure, which is so awesome in its extent that it is hard for us to comprehend.[29]

Because of strong resistance to the one-child policy in China, especially in rural areas, less emphasis is now being placed on it. It has been widely enforced in urban areas, but in rural areas, where 70 percent of the people still live, couples are usually allowed to have a second child if the first child is a female. A male child is still strongly desired in these areas to carry on the family name, to take care of his parents when they get old (an old age security system still does not exist in the rural areas), and to help with agricultural work. The one-child policy has also not been applied to ethnic groups in the country, partly because many of them live in strategic border areas and the government does not want to cause resentment among them.

China's birth control policies have been both admired and criticized in other countries. Admiration has been given for the spectacular accomplishment, for producing "one of the fastest, if not the fastest, demographic transitions in history."[30] The one-child policy has been criticized because of the means used to enforce it, which have included the use of abortions as a backup to contraceptives – sometimes on women who were strongly opposed to the procedure. Concern has also been expressed with the unnaturally low numbers of female births being reported. At first it was feared that this indicated that, in order to keep within the one-child policy and to guarantee that the couple had a male child, couples were either aborting a pregnancy if the fetus was female, or practicing female infanticide. Recent studies report that although some couples probably are using abortion for this purpose – ultrasound equipment is now widespread throughout China and ultrasound tests can be used to indicate the sex of the fetus – the underreporting of female births is probably the main cause of the so-called "missing girls" rather than female infanticide .[31]

Promoting Growth

Although most countries now seem to realize the need to limit population growth, a few recently have favored increasing their populations, among them the military governments that ruled in Argentina and Brazil in the 1960s and 1970s. Both countries have large areas that are still sparsely populated and both are rivals for the role of being the dominant power in Latin America. A few Brazilian military officers even advocated encouraging population growth so that Brazil could pass the United States in size and become the dominant nation in the Western hemisphere. It is doubtful a larger population could ever put Brazil in this position unless the economy makes great advances. A major region of the country with relatively dense population already – the northeast – is one of the poorest areas in the world and vast tracts of Brazilian land, such as in the Amazon River basin, cannot support large populations.

Aside from some pro-growth statements, the Brazilian military governments did not effectively promote population growth. They became basically neutral on the issue of population and gradually made it possible for the main nongovernment family planning organization to operate in the country. After the military left power in Brazil in the mid-1980s, a new constitution acknowledged the right of women to family planning, which the majority of Brazilian women now use. This provision had the tacit approval of the Brazilian Catholic Church.

Romania: a disastrous pro-birth policy

Romania is an example of a country that tried to promote the growth of its population. After World War II, the birth rate there fell so sharply that within a few years the population of the country would actually have started declining. In the mid-1960s the communist government, headed by Nicolae Ceausescu, decided to try to reverse this trend, not only to ward off a possible decline of population but to actually increase the number of people. Ceausescu believed that a large population would improve Romania's economic position and preserve its culture since Romania was surrounded by countries with different cultures. A great nation needs a great population, said Ceausescu. He called on all women of childbearing age to have five children. Monthly – and in some places, even weekly – gynecological exams were given to all working women 20 to 30 years old. If a woman was found to be pregnant, a "demographic command body" was called in to monitor her pregnancy to make sure she did not interrupt it. A special tax was placed on those who were childless.

The main techniques the government used to promote its pro-growth policy were to outlaw abortion, which was one of the main methods couples had used in the postwar period to limit the size of their families, and to ban the importation and sale of contraceptives. The birth rate immediately shot up, but within a few years it was nearly back to its previous low as couples found other means to limit their families. One of the means was secret abortions, and many women either died or ended up in hospitals after abortions were performed or attempted by incompetent personnel. Another tragic result of Ceausescu's pro-birth policy (as well as of his failed economic policies) was the abandoning of unwanted children. Tens of thousands of these children ended up in understaffed and ill-equipped orphanages.* Many babies were even sold for hard currency to infertile Western couples. The pro-growth policy ended in 1989 with the overthrow of the Ceausescu regime and with his execution.

* A photo essay on this subject is contained in James Nachtwey, "Romania's Lost Children," *New York Times Magazine* (June 24, 1990), pp. 28–33.

Other countries, such as Mongolia and some in sub-Saharan Africa, have at times advocated larger populations both for strategic reasons and because of the belief that a large population is necessary for economic development. Even the US government, which generally recognizes the need for a check on population growth, has some policies that promote large families, such as income tax laws that allow deductions for children. Many developed nations have contradictory policies, some encouraging population growth while others discourage it. Some developing nations also have such contradictions, although the greater agreement now in these countries about the need to limit growth often causes these contradictions to be exposed and eliminated.

A generalization one can make about governmental policies that are aimed at influencing population growth is that, aside from drastic measures, governmental policies have not been very successful in either promoting or limiting birth rates very much if these policies are out of line with what the population desires. One can also generalize that matters pertaining to reproduction are still considered to be basically the subject for private decisions, not matters for public policy to control.

The Future

The Growth of the World's Population

The United Nations projects that the world's population could continue to grow until it peaks between 8 and 28 billion, depending on the success of efforts to control population growth. The most likely total, according to the UN, is about 12 billion which could be reached around the year 2150.[32] The UN bases its projection on the assumption that the world's population growth rates will continue the decline that started in the late 1960s. In 1994 the world's population was estimated to be about 5.6 billion and it is expected to increase to about 6.25 billion by the year 2000.

The US Bureau of the Census projects that the population of the United States will continue to grow, reaching about 400 million around 2050. The US population in 1994 was estimated to be about 260 million.

The Carrying Capacity of the Earth

Will the earth be able to support a population of 12 billion or will catastrophe strike before that figure is reached? Understanding the concept of "carrying capacity" will help answer that question. Carrying capacity is the number of individuals of a certain species that can be sustained indefinitely in a particular area. Carrying capacity can change over time, making a larger or smaller population possible. Human ingenuity has greatly increased the carrying capacity of earth to support human beings, for example, by increasing the production of food. (This was unforeseen by Thomas Malthus, who wrote about the dangers of overpopulation in the late 1700s.) But carrying capacity can also change so that fewer members of the species can live. A climate change might do this. Care must be exercised when using the concept of carrying capacity because, in the past, its definition implied a balance of nature. Many ecologists no longer use the concept of balance of nature because numerous studies have shown that nature is much more often in a state of change than in a balance.[33] Populations of different forms of life on earth are usually in a state of flux as fires, windstorms, disease, changing climate, new or decreasing predators, and other forces make for changing conditions and thus changing carrying capacity.

There are four basic relationships that can exist between a growing population and the carrying capacity of the environment in which it exists. A simplified depiction of these is given in figure 2.8.

Graph (a) illustrates a continuously growing carrying capacity and population. Although human ingenuity as seen in the agricultural revolution (to be discussed in Chapter 3), and in the industrial/scientific revolution has greatly increased the capacity of the earth to support a larger number of human beings, it is doubtful the human population can continue to expand indefinitely. A basic ecological law is that the size of a population is limited by the short supply of a resource needed for survival. The scarcity of only one of the essential resources for humans – which would include air, energy, food, space, nonrenewable resources, heat, and water – would be enough to put a limit on its population growth. It is unknown how much farther the carrying capacity can be expanded before one of the limits is reached .

Graph (b) figure 2.8 illustrates a population that has stabilized somewhat below the carrying capacity. (In actuality the population may fluctuate slightly above and below the carrying capacity, but the carrying capacity remains basically unchanged.) Examples of this are seen in the undisturbed tropical rain forests where many species are relatively stable in an environment where average temperature and rainfall vary little.[34] Graph (c) portrays a situation where the population has overshot

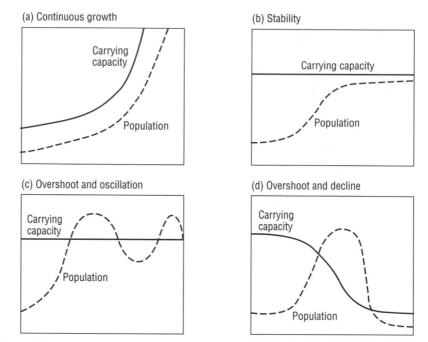

Figure 2.8 A growing population and carrying capacity.

the carrying capacity of the environment and then oscillates above and below it. An example of this situation may be the relationship between the great gray owls and their prey, lemmings and voles, in northern forests. Lemmings and voles are an important food source for the owls. Their populations rapidly increase over a period of five or six years and then, as they consume most of their food source, their numbers crash catastrophically, causing the owls to flee the area to escape mass starvation.[35] Graph (d) illustrates a situation in which the overshooting of the carrying capacity leads to a precipitous decline in the population, or even to its extinction, and also to a decline in the carrying capacity. Such a situation has occurred with deer on the north rim of the USA's Grand Canyon[36] and with elephants in Kenya's Tsavo National Park.[37] In both cases, the number of animals increased to a point where they destroyed the vegetation they fed upon.

It is my hope that the human species with its unique mental powers will create a situation that combines elements of graphs (a) and (b), using its abilities to increase the carrying capacity of earth, where possible, and where not, making sure its numbers do not exceed that capacity. But there are many indications that the species has not yet recognized its danger and is not yet taking effective efforts to prevent either situation (c) – which would mean the death of millions – or situation (d), which could lead to the decline of the human race. There are places in the world where population expansion has already passed the carrying capacity of the land and the land itself is now being destroyed; in sub-Saharan Africa, for example, fertile land is turning into desert and in the Himalayan mountain area, land is being destroyed by human-made erosion and floods. There are many other examples of the reduction of the carrying capacity of the earth that is taking place at unprecedented rates today around the world – the result of uncontrolled over-grazing, over-fishing, over-planting, over-cutting of forests, and the over-production of waste leading to pollution. (Some of this reduction of carrying capacity is being caused by population pressures and some by economic forces, e.g., the desire to increase short-term profits.) This deterioration has led many ecologists to believe that unless there is a rapid and dramatic change in many governmental policies, the human species may indeed be headed for the situations depicted in either the oscillation or decline graphs above.

Optimum Size of the Earth's Population

What is the optimum size of the earth's population? That question, like others we have asked in this chapter, is not going to be easy to answer, but it is worth asking. Paul Ehrlich, professor of population studies and of biology at Stanford University, defines the optimum size of the earth's population as that "below which well-being per person is increased by further growth and above which well-being per person is decreased by further growth." What does "well-being" mean? Ehrlich explains what he believes it means:

> The physical necessities – food, water, clothing, shelter, a healthful environment – are indispensable ingredients of well-being. A population too large and too poor to be supplied adequately with them has exceeded the optimum, regardless of whatever other aspects of well-being might, in theory, be enhanced by further growth. Similarly, a population so large that it can be supplied with physical necessities only by the rapid consumption of nonrenewable resources or by activities that irreversibly degrade the environment has also exceeded the optimum, for it is reducing Earth's carrying capacity for future generations.[38]

Ehrlich believes that, given the present patterns of human behavior – behavior that includes the grossly unequal distribution of essential commodities such as food and the misuse of the environment – and the present level of technology, we have already passed the optimum size of population for this planet. Interestingly, government officials in China may have come to a similar conclusion regarding their own country. The official Chinese press in 1990 mentioned that 700 million may be the ideal population for the country – a reduction of about 400 million from the country's population at that time.[39]

Julian Simon, professor of business administration at the University of Maryland and author of *The Ultimate Resource*, believes that the ultimate resource on earth is the human mind. The more human minds there are, says Simon, the more solutions there will be to human problems. Simon admits that population growth in the less developed countries can lead to short-term problems since more children will have to be fed. But in the long run these children will become producers, so the earth will benefit from their presence. Simon agrees that rapid population growth can harm development prospects in poor nations, but he is not disturbed by moderate growth in these countries. Larger populations make economies of scale possible; cheaper products can be made if there are many potential consumers. Also, services can improve, as seen by the development of efficient mass transportation in Japan and Europe in areas of dense population.

Simon's views won favor in the Reagan and Bush administrations and were used to give academic support to a new US policy on population – popularly called the Mexico City policy. This policy, which has been discussed earlier in this chapter, basically saw the effect of population growth as a "neutral phenomenon ... not necessarily good or ill," a position that Marxist ideology also held.[40] While many economists in the United States do not share Simon's view that "more is better," many do share his view that human ingenuity, especially new technology and resource management practices, can increase the carrying capacity of the earth as it has in the past.[41]

Population Problems in Our Future

Throughout this book we are going to be looking at the many current problems related to overpopulation and its causes. Here I will mention a few of the most important ones. Hunger is an obvious problem in which overpopulation plays a key role, and the number of hungry people is undoubtedly increasing. The news media are used to dramatizing this problem only when there are many children with bloated bellies to be photographed, but much more common than the starving child today, and probably in the future, will be the child or adult who is permanently debilitated or who dies because of malnutrition-related diseases. Pollution and the depletion of nonrenewable resources will increase as the world's population grows. Migration of people to lands that do not want them will probably increase in the future and this will cause international tension. An estimated 3 million immigrants entered the United States illegally during the 1980s, many of them Mexicans looking for work.[42]

In Assam, India, several thousand unwanted immigrants were massacred in 1983. Wars have taken place in the past in which overpopulation played an important role and they will probably occur in the future. In the 1960s a border war broke out between El Salvador and Honduras over unwanted Salvadorians in Honduras. In the early 1990s numerous brutal civil wars were occurring in Africa. While we cannot identify overpopulation as the main cause of these conflicts, it is likely that increasing population pressures made the ethnic conflicts more likely.

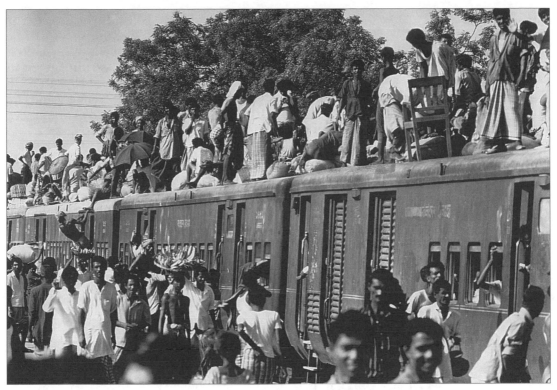

A more frequent picture in the future? A crowded train in Bangladesh. *(World Bank)*

And growing populations in countries situated in regions with serious water shortages – for example, in the Middle East – are a direct cause of competition and conflict over the scarce water. It has been estimated that by the year 2025 about 1 billion people in Africa and southern Asia will have scarce water supplies.[43] Millions of people in North Africa and in the Middle East already live in conditions of serious water shortages. In some of these regions droughts have been common throughout history. What is not common in these regions is the population density that is present and projected. While water scarcity can obviously promote conflict, it also has the potential for promoting cooperation as nations are forced to devise ways to conserve and share scarce water .

Population pressure might even ultimately lead to the decline of our civilization. Recent research into the mysterious collapse of the Mayan civilization in Central America suggests that an exponentially growing population may have put pressures on the environment which led to the collapse.[44]

Conclusions

How should I end this chapter? As I look over the statements by a number of population experts, they seem to share the conclusion that the earth faces an overwhelming problem with its current population growth of about 1 million people every four days. This is a problem that is second only to the threat of nuclear weapons for having the potential for causing untold human misery. But many of these experts also emphasize that human thinking and governmental policies are starting to change and some impressive reductions in birth rates are taking place in a few countries. We know how to reduce birth rates. What is lacking at present is the political will to do what needs to be done to address the problem. We will do it if we take seriously the warning given in 1991 in a joint statement by the US National Academy of Sciences and its British counterpart, the Royal Society of London:

> If current predictions of population growth prove accurate and patterns of human activity on the planet remain unchanged, science and technology may not be able to prevent either irreversible degradation of the environment or continued poverty for much of the world.[45]

NOTES

1 Susan Kalish, "New UN Projections Include Local Effects of AIDS," *Population Today,* 20 (Washington: Population Reference Bureau, October 1992), p. 1.

2 Malin Falkenmark and Carl Widstrand, "Population and Water Resources:A Delicate Balance," *Population Bulletin,* 47 (Washington: Population Reference Bureau, November 1992), p. 23.

3 Edward Gargan, "On Meanest of Streets, Salvaging Useful Lives," *New York Times,* national edn (January 8, 1992), p. A2.

4 The United Nations uses the term "urban agglomeration" in place of "city." An urban agglomeration covers a much larger area than that of the administrative boundaries of the central city. It includes the central city and the continuous, densely populated area surrounding the city. It is the urban area as if seen from an airplane and is generally recognized to be a more useful concept than the idea that a "city" is only that area within the administrative boundaries. My figure uses the UN's broader definition of what is a city.

5 Lester R. Brown, *The Twenty-ninth Day* (New York: W. W. Norton, 1978), p. 73.

6 Cited in Paul R. Ehrlich, Anne H. Ehrlich, John P. Holdren, *Ecoscience: Population, Resources, Environment* (San Francisco: W.H. Freeman, 1977), pp. 777–8.

7 John C. Caldwell and Pat Caldwell, "High Fertility in Sub-Saharan Africa," *Scientific American,* 262 (May 1990), pp. 118–25.

8 Bryant Robey, Shea O. Rutstein and Leo Morris, "The Fertility Decline in Developing Countries," *Scientific American,* 269 (December 1993), p. 67.

9 Brown, *Twenty-ninth Day,* p. 77.

10 National Research Council, *Population Growth and Economic Development: Policy Questions* (Washington: National Academy Press, 1986, p. 90.

11 Machiko Yanagishita, " Japan's Declining Fertility: '1.53 Shock'" *Population Today*, 20 (Washington: Population Reference Bureau, April 1992), pp. 3–4.

12 William Schmidt, "Retirement Resort in England Is a Model for Europe," *New York Times*, national edn (July 13, 1993), p. A1.

13 Joseph A. McFalls, Jr "Population: A Lively Introduction," *Population Bulletin*, 46 (Washington: Population Reference Bureau, October 1991), p. 36.

14 A discussion of the political forces that were instrumental in the Reagan and Bush administrations in developing this policy and in keeping it in force for ten years is contained in Michael S. Teitelbaum, "The Population Threat," *Foreign Affairs,* (Winter 1992/1993), pp. 63–78. The policy was reversed in 1994 after the Clinton administration took office.

15 Alan Cowell, "U.N. Population Meeting Adopts Program of Action," *New York Times*, national edn (September 14, 1994), p. A2.

16 Robey et al., "The Fertility Decline," p. 63.

17 Ibid., pp. 60–7. Bangladesh is cited as an example of an extremely poor country where the educational level of women is very low (three out of four women are still illiterate) but which has had a significant decline in fertility because of the effectiveness of its family planning program. Yet birth rates are still high in Bangladesh and some experts believe that for a further decline to take place, economic development must occur and the social and economic status of women must improve. For example, see Susan Kalish, "Culturally Sensitive Family Planning: Bangladesh Story Suggests It Can Reduce Family Size," *Population Today*, 22 (Washington: Population Reference Bureau, February 1994), p. 5.

18 Robey et al., "The Fertility Decline," p. 65.

19 Ibid., p. 60.

20 *New York Times,* late city edn (August 18, 1974), p. 2.

21 Peter J. Donaldson and Amy Ong Tsui, "The International Family Planning Movement," *Population Bulletin,* 45 (Washington: Population Reference Bureau, November 1990), pp. 4–5.

22 The figure 75 percent is highly significant because it is the level of contraceptive use that is statistically associated with a two-child family average, what demographers call "replacement level fertility."

23 Donaldson and Tsui, "The International Family Planning Movement," p. 20.

24 Robey et al., "The Fertility Decline," p. 67.

25 Population Reference Bureau, *1994 World Population Data Sheet* (Washington: Population Reference Bureau, 1994).

26 "Mexican Men Get the Message About Limiting Family Size," *Christian Science Monitor* (July 8, 1992), p. 11.

27 Ajoa Yeboah-Afari, "Male Responsibility: Still a Missing Link," *Popline* 13, (Washington: Population Institute, May-June 1991), p. 6.

28 "Mexican Men Get the Message About Limiting Family Size," *Christian Science Monitor* (July 8, 1992), p. 11.

29 Ronald Freedman, "Theories of Fertility Decline: A Reappraisal," In Philip M. Hauser (ed.), *World Population and Development* (Syracuse, N.Y.: Syracuse University. Press, 1979), p. 75.

30 H. Yuan Tien et al., "China's Demographic Dilemmas," *Population Bulletin,* 47 (Washington: Population Reference Bureau, June 1992), p. 6.

31 See, for example, "China's Sex Ratio," *Population Today,* 21 (Washington: Population Reference Bureau, December 1993), p. 8.

32 For an analysis suggesting that basically we do not know what the future size of the world's population will be, see Carl Haub, "New UN Projections Show Uncertainty of Future World," *Population Today,* 20 (Washington: Population Reference Bureau, February 1992), pp. 6–7.

33 Daniel B. Botkin, "A New Balance of Nature," *Wilson Quarterly* (Spring 1991), pp. 61–72; and William K. Stevens, "New Eye on Nature: The Real Constant is Eternal Turmoil," *New York Times,* national edn (July 31, 1990), pp. B5–B6.

34 G. Tyler Miller Jr, *Living in the Environment: Principles, Connections, and Solutions,* 8th edn (Belmont, Calif.: Wadsworth, 1994), p. 153.

35 David Attenborough, *The Living Planet: A Portrait of the Earth* (Boston: Little, Brown, 1984), p. 76.

36 Edward J. Kormondy, *Concepts of Ecology,* 2nd edn (Englewood Cliffs, NJ: Prentice-Hall, 1976), p. 111–12.

37 Daniel Botkin, "A New Balance of Nature," pp. 61–3.

38 Ehrlich et al., *Ecoscience,* p. 716.

39 Nicholas Kristof, "More in China Willingly Rear One Child," *New York Times,* national edn (May 9, 1990), pp. A1 and B9.

40 Michael Teitelbaum, "The Population Threat," pp. 71–2.

41 An interesting wager that Ehrlich and Simon made about whether the world's growing population was running out of natural resources (and which was won by Simon) is reported in John Tierney, "Betting the Planet," *New York Times Magazine,* (December 2, 1990), pp. 52–81.

42 "Movement Across Borders Underestimated," *Popline,* 15 (Washington: Population Institute, July – August 1993), p. 7.

43 Falkenmark and Widstrand, " Population and Water Resources," p. 20.

44 E. S. Deevey et al., "Mayan Urbanism: Impact on a Tropical Karst Environment," *Science,* 206 (October 19, 1979), pp. 298–306; and Don S. Rice , "Roots: Resourceful Maya Farmers Enabled a Mounting Population to Survive in a Fragile Tropical Habitat," *Natural History* (February 1991), pp. 10–14.

45 "A Warning on Population," *Christian Science Monitor,* (March 24, 1992), p. 20.

Further Readings

Critchfield, Richard, *The Villagers: Changed Values, Altered Lives: The Closing of the Rural-Urban Gap* (Garden City, N.Y.: Anchor Press/Doubleday, 1994). Soon, for the first time in human history, more people in the world will be living in urban rather than rural areas. What will be lost? Critchfield believes a lot since religious, family, and ethical values have their base in rural life.

Dixon-Mueller, Ruth, *Population Policy and Women's Rights: Transforming Reproductive Choice* (Westport, Conn.: Praeger, 1993). The author explains why she believes that efforts to control fertility in the less developed countries must become more responsive to the concerns of the women in those countries. Population policies must become more centered on the issues of women's survival and security.

Donaldson, Peter J., and Amy Ong Tsui, "The International Family Planning Movement," *Population Bulletin,* 45 (Washington: Population Reference Bureau, November 1990). A detailed examination of the evolution of the international family planning movement and an evaluation of its effectiveness.

Eberstadt, Nicholas, "Population Change and National Security," *Foreign Affairs* 70 (Summer 1991), pp. 115–31. Concerned with the effect of changes in global demographics on the national security of the United States, Eberstadt explores the forces that shape population, such as fertility, mortality, health care, and economics in the Third World.

Ehrlich, Paul R. and Anne H. Ehrlich, *The Population Explosion* (New York: Simon and Schuster, 1990). Taking the neo-Malthusian position that the world is heading for disaster because of overpopulation, the Ehrlichs argue that the earth's population is already too large for the earth's ecosystem. Continuing the argument they first made two decades ago in the *Population Bomb,* they state that unless major efforts are made to rein in our numbers, natural forces (such as widespread death) will soon bring the human population back into balance.

Fornos, Werner, *Gaining People, Losing Ground.* (Ephrate, Pennsylvania: Population Institute, 1987). A highly readable discussion of the pressures of population on society and the environment, including case studies of several Third World countries.

Hardin, Garrett, *Living Within Limits: Ecology, Economics, and Population Taboos* (New York: Oxford University Press, 1993). Hardin argues that we are heading for an ecological catastrophe unless Draconian steps are taken to stop the growth of human population. He believes that each nation must take care of its own population problem and that some form of governmental coercion is needed to address this problem.

Jacobson, Jodi L., "Improving Women's Reproductive Health," *State of the World* 1992, (New York: W. W. Norton, 1992), pp. 83–99. Motherhood is a dangerous occupation in the Third World. This article presents ways this danger can be reduced.

James, Valentine (ed.), *Urban and Rural Development in Third World Countries* (Jefferson, NC: McFarland, 1991). Clear and understandable text highlights an in-depth exploration of the relationship between population, culture, and technology in Third World development.

Keyfitz, Nathan, "The Growing Human Population," *Scientific American,* 261 (September 1989), pp. 119–26. A general, nontechnical article about the significance of the population growth taking place on earth at present and of efforts to control it.

Livi-Bacci, Massimo, *A Concise History of World Population* (Cambridge, Mass: Blackwell Publishers, 1992). A readable history of the growth of the human population from prehistoric times to the present. The author takes a position in between that of the optimists and those predicting a catastrophe; he presents an alternative way to understand and deal with population growth.

Lutz, Wolfgang, "The Future of World Population," *Population Bulletin,* 49 (Washington: Population Reference Bureau, June 1994). The only certainties that exist regarding world population are that world population will continue to grow, the proportion living in developing countries will continue to increase, and the average age of the population in all regions will rise. Aside from these three certainties, there is a wide range of possibilities regarding the future size of world population.

McFalls, Joseph A., Jr, "Population: A Lively Introduction," *Population Bulletin,* 46 (Washington: Population Reference Bureau, 1991). The author explains the forces that cause populations to grow or decline, and that determine the age and sex distribution within a society. McFalls summarizes the unprecedented population growth that has taken place in the world over the past few centuries and looks at the most important population-related issues facing the world at present.

Simon, Julian L. *Population Matters: People, Resources, Environment, and Immigration* (New Brunswick, NJ: Transaction Publishers, 1990). This challenge to the doomsayers' vision of food shortages, environmental damage, and economic disaster looks at population growth from a positive viewpoint. Disagreeing with many social scientists and ecologists about the population issue, Simon believes that population growth has long-term benefits and that the human condition is much better than the neo-Malthusians admit.

3

FOOD

One way a civilization can be judged is by its success in reducing suffering. Development can also be judged in this way. Is it reducing the misery that exists in the world? Throughout human history, hunger has caused untold suffering. Because food is a basic necessity, when it is absent or scarce humans need to spend most of their efforts trying to obtain it; if they are not successful in finding adequate food, they suffer, and, of course, can eventually die. In the United States, where dieting is a major preoccupation, there is a tendency to ignore the problem of hunger. It is so far removed from the daily experience of most US citizens that they tend to forget it is a problem for many in the world. In this chapter we will look at hunger and also at a problem the more developed countries face: how their own level of development affects the food they eat.

Food Production ● ● ● ● ● ● ● ● ● ● ● ● ● ● ● ● ● ●

How much food is produced in the world at present? Is there enough for everyone? The answer, which may surprise you, is that, yes, there is enough. There is enough grain grown in the world today to provide every man, woman, and child with 3,000 calories a day, which is only a little less than that consumed by the average person in the industrialized countries and well above the minimum requirement for humans.[1] (A proper diet must include more than just grain, of course, but throughout the world grain is an important part of the human diet.)

In the 1970s and 1980s the world's output of major food crops increased significantly – the most dramatic increase evident in the production of cereals, which rose about 50 percent – as improved seeds, machinery, fertilizers, and pesticides were used to increase production and new land was cultivated. (Most of this growth in production came from an increase in yield per acre rather than from an increase in the amount of cropland.[2]) This impressive performance was counterbalanced, however, by the rapid growth of population that was also taking place in the world at this time. But food production increased rapidly enough in the 1970s and 1980s so that, except for one region – Africa – the output of food in the world kept up with population growth. The best performance was in the centrally planned economies of Asia where production stayed well ahead of population growth. In the Middle East and in Latin America production barely managed to stay even with population growth. It was only in Africa that there was a decline in the per capita food output in these two decades because of poor performance in agriculture (which was caused in part by droughts, civil wars, and non-supportive government policies) and because of very rapid population growth.[3]

How Many Are Hungry?● ● ● ● ● ● ● ● ● ● ● ● ● ●

Unprecedented amounts of food in the world do not mean, unfortunately, that everyone is getting enough food. No one knows for sure how widespread hunger[4] is today. Estimates vary from about 500 million hungry to about 1 billion.[5] With spreading economic development, famines are becoming rarer than they were in the past. But several major famines have occurred in the twentieth century. In the Soviet Union in the early 1930s Stalin forcibly collectivized agriculture and deliberately caused a famine in the area where most of the grain was grown – the Ukraine and Northern Caucasus – in order to break the resistance of the peasants. An estimated seven million people – three million of them children – died in that famine.[6] Another country with a communist government experienced the worst famine in the twentieth century. Although it was kept secret from the outside world while it was occurring, China had a famine in the late 1950s and early 1960s which led to an estimated 16 to 64 million deaths. The famine was caused by both mistaken governmental actions and bad weather.[7] Except for Africa, actual starvation is uncommon in the present world. A much larger number of people die today because of malnutrition, a malnutrition that weakens them and makes them susceptible to many diseases. Children die from diarrhea in poor countries – a situation nearly unheard of in rich countries – partly because of their weakened condition.

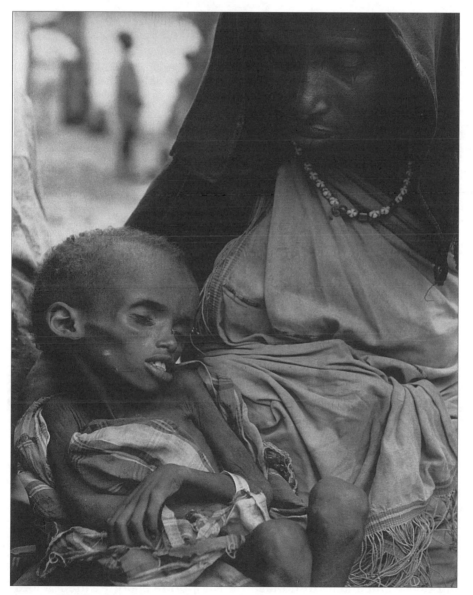

Starvation in Somalia. (CARE: Zed Nelson)

Who are the hungry and where do they live? The answer to the first question is that, according to World Bank estimates, 80 percent are women and children. Malnutrition is common among the poor in many developing countries, and most of these people live in small villages. Because the men are needed to acquire food for the family, they are fed the best. One private US organization working to alleviate world hunger has estimated that about 40 percent of the hungry are landless tenant farmers and agricultural workers, another 40 percent are subsistence farmers, and the final 20 percent live in urban slums.[8]

What the poor eat in Calcutta

Here is what some of the poor eat in Calcutta, as described by a US journalist who was invited to share a meal:

Ahmed...invited me for tea at the kiosk, where he shined shoes. A monsoon storm broke, and about a dozen of us ended up spending an hour or two talking together. All of them were either shoeshine men, beggars, or pickpockets. Chowringhee Road was their universe. A meal was served in a big tiffin (luncheon) can. It was rice and curry mixed together – leftovers scraped off the plates at a nearby government canteen by some enterprising Bengali. One portion cost about five cents. It was dumped in a big pile on a newspaper and everyone squatted around in a circle, avidly eating with his fingers.

Source: Richard Critchfield, *Villages* (Garden City, NY: Anchor Press/Doubleday, 1981), pp. 285–6.

About one-half of what the World Bank has called the "absolute poor" (those with so little money, goods, and hope to put them in a special class) live in India, Bangladesh, and Pakistan. A large number of them live in Indonesia and in sub-Saharan Africa. The rest are scattered throughout the Middle East, Latin America, and the Caribbean.[9]

There are indications that the number of hunger-related deaths in the world has decreased during the past 30 years in spite of the world's growing population. It was estimated in the mid-1980s that about 15 million people were dying each year at that time from hunger-related causes.[10]

Causes of World Hunger

If there is more than enough food being grown at present for the world's population but up to a quarter of the earth's people are malnourished, what is causing hunger in the world? Food authorities generally agree that poverty is the main cause of world hunger. Millions of people do not have enough money to buy as much food as they need, or better kinds of food. This is the reason one food expert has written that "Malnutrition and starvation continue more or less unchanged through periods of world food glut and food shortage."[11] The world's poorest cannot afford to purchase the food they need, whatever its price.

Other low-income people in the Third World suffer during food shortages when the price of food increases dramatically, as it did during the early 1970s when world prices of rice, wheat, and corn doubled in just two years. The poor traditionally spend 60 to 80 percent of their income on food. If world demand is high for certain foods, such as beef for the US fast-food market, then the large landowners in Third World countries grow food or raise cattle for export rather than for domestic consumption. This tends to cause domestic food prices to increase since the supply of local foods is reduced. Much land was being used to grow export crops such as cotton and peanuts in the Sahel, a huge area in Africa just south of the Sahara desert, when a famine hit that area in the early 1970s. Six years of drought, rapid population growth, and misuse of the land led to widespread crop failures and livestock deaths. It is estimated that between 200,000 to 300,000 people starved to death in the Sahel and in Ethiopia before international aid reached them.[12]

Famine also hit Cambodia in the 1970s. Years of international and civil war, coupled with the genocidal policies of the communist government under the leadership of Pol Pot, led to an estimated 10,000 to 15,000 people dying every day during the worst of the famine in 1979. A highly successful international aid effort, first organized by private organizations and then joined by governmental agencies, saved the Cambodian people from being destroyed.

Famine hit again in Africa in the mid-1980s and early 1990s. Televised pictures of starving people in Ethiopia led to a large international effort by private organizations and by governments to provide food aid. The famine in Ethiopia, Somalia, and the Sudan and in other sub-Saharan African countries was not caused only by the return of a serious drought to the region. Many of the causes of these famines were the same as those that brought on the famine in Africa in the early 1970s. In addition to the reasons stated above, the extensive poverty in the region, a world-wide recession that seriously hurt the export-oriented economies of the African countries, civil wars, and governmental development policies that placed a low priority on agriculture have been identified as likely causes.[13]

It is easier to pinpoint the reasons for the famines than for the global malnutrition that exists. Let's look at Bangladesh, which has been called an "international basket case" by some commentators. Its situation is indeed desperate. In a country about the size of the state of Wisconsin (which has a population of 5 million), live about 100 million people. It is one of the most densely populated and one of the poorest countries on earth, and it has one of the highest birth rates. Nutritionists state that the average consumption of food is barely above the starvation level.[14] Yet its land is green and lush and Bangladesh has some of the most fertile soil on the planet. It has plenty of water, and aside from not infrequent devastating storms, has a climate that can support three harvests a year. According to some studies, there is probably enough food grown there now to feed the present population adequately, and enough could be grown even to provide for the large increase in population that is expected to take place there in the next 20 years.

So what is causing the hunger in Bangladesh? According to the above studies, the feudal economic and social systems are mainly to blame. There is a very unequal distribution of land, with the wealthiest 17 percent of the rural population owning two-thirds of the land. Nearly 60 percent of the rural population owns less than one acre of land per person, and, with the rapid population growth, the number who own no land at all is growing. The average rural wage is about 50 cents a day. Absentee landlordism is common, and since landlords have a major influence in the government, it is unlikely that serious land reform measures will be passed.[15]

According to a study on world hunger that went to the US president in 1980, the developing and the developed nations must share the blame for allowing world hunger to exist. Here is what the report of the Presidential Commission said about the responsibility of the developing nations:

> In the developing countries, domestic political problems, national security questions, and industrial development generally have attracted more attention and resources than alleviating poverty or investing in agriculture. In fact, few nations have even made these latter concerns top developmental priorities. The development of remote, backward rural areas has had little political or psychological appeal to civilian or military rulers bent on maintaining their control and modernizing their societies along sophisticated technological and industrial lines. Adopting the priorities of the industrialized world, many Third World leaders have modernized their armies and parts of their cities at the expense of their agriculture, health care, and education.

The Commission stated that the United States and other developed nations must share the responsibility for world hunger because their foreign aid policies have been focused in other directions.

> The United States and other developed nations, too, have placed a low priority on alleviating world hunger. Since World War II, the industrialized countries have been preoccupied with East-West tensions and sustaining domestic economic growth. These primary concerns have largely determined both the nature and extent of the West's involvement with the developing world. With national security and anti-Communism as paramount concerns, more money has always been available for military assistance, arms transfers, and the training of military personnel than for educating teachers, scientists, economists, farmers, and health care specialists... .The hard reality is that the overwhelming majority of the world's hungry people live in

countries which have been of limited significance to world grain markets and to Western geopolitical concerns.[16]

The Commission concluded that, "the issue of ending world hunger comes down to a question of political choice... .The Commission agrees with other studies that, if the appropriate political choices were made, the world can overcome the worst aspects of hunger and malnutrition by the year 2000."[17]

How Food Affects Development ● ● ● ● ● ● ● ● ● ● ● ● ● ● ● ● ●

The availability of food has a direct effect on a country's development. Possibly the most destructive and long-lasting is the effect that the absence of food – or, more often, of the right kinds of food – has on the children of the less developed nations. As mentioned in chapter 2, the dying of many children in poor nations at birth or in their first few years is one of the causes of high birth rates. The United Nations estimates that about one-half of the children under five in the typical less developed nation suffer from some form of malnutrition.[18] A deficiency of vitamin A leads to blindness of about 100,000 children a year in developing countries.[19] More common than blindness are the harmful effects malnutrition has on the mental development of the children. Eighty percent of the development of the human brain occurs before birth and during the first two years after birth. Malnutrition of the pregnant mother or of the child after birth can adversely affect the child's brain development and, along with limited mental stimulation, which is common in poor homes, can lead to a reduced capacity for learning.

The bloated belly is a sign of malnutrition, a major cause of stunting and death in the children of the developing world.
(CARE: Joel Chiziane)

Malnutrition also reduces a person's ability to ward off diseases since it reduces the body's natural resistance to infection. Measles and diarrhea, which are generally nonserious illnesses in the developed nations, often lead to the death of children in the developing nations; in fact, diarrhea is the single greatest cause of death of children in the Third World. When a child has been weakened by malnutrition, sickness is likely to come more frequently and to be more serious than that experienced by the well-nourished child.

Some US citizens, when they first visit the Third World, come away with a feeling that the people are lazy since they are likely to see a number of people sitting around, not doing much of anything. Aside from the absence of jobs or of land they can farm, malnutrition also may be playing a role here because, as one study has stated, "Chronically undernourished people, who commonly also suffer from parasitism and disease, are typically apathetic, listless, and unproductive."[20]

The presence of unhealthy and unproductive people in rural areas probably means that not as much food is being grown as is possible, and the presence of unhealthy and unproductive people in urban areas probably means that not as many manufactured goods and services are being produced. A nation that must spend scarce foreign exchange to buy imported food cannot use those funds to support its development plans. And, more importantly, a nation whose main and most important resource – its people – is weakened by malnutrition is unlikely to generate the kind of economic development which actually does lead to an improved life for the majority of its people. James Grant, the head of the United Nations Children's Fund (UNICEF), has described well the interrelatedness of all key elements of development:

> A cat's cradle of... synergisms links almost every aspect of development: female literacy catalyzes family planning programmes; less frequent pregnancies improves maternal and child health; improved health makes the most of preschool or primary education; education can increase incomes and agricultural productivity; better incomes or better food reduces infant mortality; fewer child deaths tend to lead to fewer births; smaller families improve maternal health; healthy mothers have healthier babies; healthier babies demand more attention; stimulation helps mental growth; more alert children do better at school...and so it continues in an endless pattern of either mutually reinforcing or mutually retarding relations which can minimize or multiply the benefits of any given input.[21]

How Development Affects Food

The development that took place in Europe and the United States as they industrialized led to an increase in the average family's income, and this meant more money to buy food. As we saw in the preceding section, poverty is the main cause of malnutrition. As incomes rose in the West, hunger disappeared as a concern for the average person. Except for some subgroups in Western countries, malnutrition is no longer a common problem.

Development also affects food in other ways. As a nation develops, major changes start to take place in its agriculture. We will look first at how development affects the amount of food that is produced and how it is produced, and then at the way development affects the types of food people eat.

The Production of Food

Western agriculture produces an impressive amount of food. The US supermarket, better than any other institution, illustrates the abundance that modern agriculture can produce. The United States produces so much food that huge amounts of important crops such as corn, wheat, and soybeans are exported. Much of this US abundance has come since the end of World War II. Since 1950, US food production has increased by 50 percent despite a decrease in the land under cultivation.[22] Table 3.1 shows that the average US farmer in the early 1990s was producing enough food for about 100 people. What is the reason for this increase in production? There are many reasons, of course, but basically, it is because American agriculture has become mechanized and scientific. By using new seeds that can benefit from generous amounts of fertilizer, pesticides, heavy machinery, and irrigation, production has soared. But this accomplishment has had its costs, as we shall see below.

Table 3.1 Number of people for whom food is produced by each US farm worker

Year	Number of people
1930	10
1940	11
1950	15
1957	23
1981	78
1992	101

Source: Data from Wayne Rasmussen and Paul Stone, "Toward a Third Agricultural Revolution," *Food Policy and Farm Programs: Proceedings of the Academy of Political Science*, 34 (1982), p. 183; and Scott Pendleton, "Family Farms Struggle, Survive," *Christian Science Monitor* (November 12, 1992), p. 9.

Western agriculture basically turns fossil fuel into food. This type of agriculture was developed when oil was inexpensive. Large amounts of energy are needed to build and operate the farm machinery, to build and operate the irrigation systems, to create the pesticides, and to mine and manufacture the fertilizers. Also, huge amounts of energy are needed to process the foods, to transport them to market, to package them, and to display them in retail stores. (Even in this period of greatly increased energy prices, the open freezer in US supermarkets is still common.) It has been estimated that to raise the rest of the world's diet to the American level – especially one matching its high consumption of beef – would consume nearly all the world's known reserves of oil in 15 years.[23] Figure 3.1 shows the amount of energy that is needed to produce various foods, and illustrates well the different amounts of energy expended in modern high-yield agriculture in comparison with traditional agriculture and with preagriculture . The figure shows that important modern foods, such as ocean fish and much of our beef and eggs, actually require an investment of more energy to produce the food than is obtained through the food.

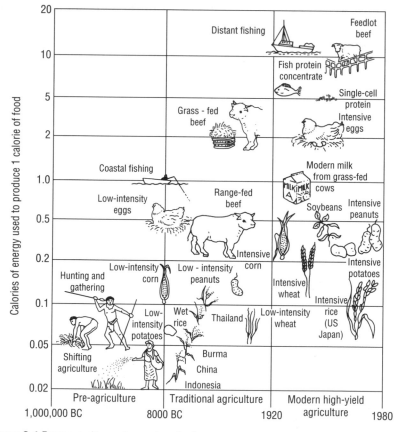

Figure 3.1 Energy used to produce various foods
Source: Adapted from Paul R. Ehrlich et al., *Ecoscience: Population, Resources, Environment,* p. 349. Copyright © 1970, 1972, 1977 W.H Freeman and Company. Used by permission.

Although modern, mechanized agriculture is generally – but not always – much more productive than the traditional agriculture more commonly found in Third World countries, traditional agriculture is generally far more energy-efficient than Western agriculture. In traditional agriculture the amount of energy used in the form of farm labor and materials is typically small compared with the yield. Returns up to 50 to 1 are possible, although more common are 15 to 1 returns, whereas in modern industrial agriculture more energy is expended than produced.[24] To produce and deliver to a US consumer one can of corn that has 270 calories in it, a total of about 2,800 calories of energy must be used. To produce about 4 ounces of beefsteak, which also provides about 270 calories, an astounding 22,000 calories of energy must be expended.[25] Anthropologists Peter Farb and George Armelagos give us one perspective we need in order to judge the effects that development, as achieved in the West, is having on agriculture:

In short, present-day agriculture is much less efficient than traditional irrigation methods that have been used by Asians, among others, in this century and by Mayans, Mesopotamians, Egyptians, and Chinese in antiquity. The primary advantage of a mechanized agriculture is that it requires the participation of fewer farmers, but for that the price paid in machines, fossil fuels, and other expenditures of energy is enormous.[26]

As the late Barbara Ward, a respected British author of many books on development, noted, "the high-energy [US] food system is one reason why the United States, for 5 percent of the world's people, is now consuming nearly 40 percent of its nonrenewable resources."[27] That statement, more than any other, presents the main argument of those who maintain that there is no way the rest of the world can adopt the agricultural methods followed by the United States at present.

Another feature of US agriculture is an increase in the size of farms and a reduction in their number. Table 3.2 shows how farm size and numbers have changed from 1940 to 1990.

Increased demand for farm products, along with government price supports, enabled farmers to replace old sources of power (horses and mules) with new sources (first the steam engine and then the gasoline engine) and to begin using more machinery, improved seeds, fertilizers, and chemicals to control pests. Because of dramatic increases in farm productivity, by 1990 only about 2 percent of US citizens were farmers, down from about 30 percent in 1920.[28] With the increasing financial investment necessary to support the new type of agriculture, and the competition the large farms provide, there has been a noticeable decline in the small, family-owned farm in the United States.

The growth of what has become known as "agribusiness" – farms run like a big business – has meant an increased concentration of control over the production of food in the United States, although there is still

Table 3.2 Number and size of US farms

	Number of farms	Average size of farms (acres)
1940	6,400,000	170
1950	5,600,000	210
1960	4,000,000	300
1970	2,900,000	370
1980	2,400,000	430
1990	2,100,000	460

Source: Data from *Statistical Abstract of the United States* (Washington: US Bureau of the Census, 1970 and 1992), p. 582 (1970) and p. 644 (1992).

sufficient competition in agriculture so that food remains relatively inexpensive. The large industrial farms can produce harvests of 100 million tomatoes, but sometimes with less efficiency than small operators can obtain. When committees make decisions instead of the farmer growing the crop, when there is inattention to detail, and when there is a lack of dedication – dedication that usually comes only when someone has a personal stake in the farm – one often finds waste and mismanagement. This happened on large state-owned farms in the Soviet Union, and it is happening on large industrial farms in the United States.

Mismanagement on an Industrial Farm: The Case of the Oversized Carrots

When the author of a book on three different types of farms in the United States saw an entire crop of carrots being plowed under instead of being harvested on a corporation-owned farm in California, he was given the following explanation by a farm supervisor:

There are enough carrots on [sic] the world right now without these... Price isn't so hot, and the warehouses were full when these got to the right size. We were held off harvesting. Someone let time go by and suddenly they were too big. More than eighty acres of them, which comes to sixty million carrots or so. They couldn't fit into those plastic carrot sacks they sell carrots in unless they were cut, and that would have cost the processor a bundle. They offered us $125 an acre for the carrots – and it would have run us $200 just to have them contract-harvested. So this is the cheapest alternative... .

Source: Mark Kramer, *Three Farms: Making Milk, Meat and Money from the American Soil* (Boston: Little, Brown, 1980), p. 248.

An abundance of food in developed nations seems to lead to increased waste. It has been estimated by a congressional report that, each year the United States throws away enough food to feed about 50 million people.[29]

Besides the waste of food, there is another waste occurring in the United States that could affect profoundly its ability to produce food in the future: the loss of its farmland. About 3 million acres of farmland in the United States is being lost annually because of development; it is being covered over by houses, roads, shopping centers, and factories, and by general urban sprawl. While the amount lost is small compared to the amount of actual and potential cropland in the United States (about 0.5 percent), the land lost is often prime farmland, and it can be replaced only by marginal land, which is not as fertile, is more open to erosion, and is more costly to use.[30] Other land is being lost because of overplanting and erosion. The planting, year after year, of certain export crops drains the soil of valuable nutrients. Also, farmers in the West started to plow up fragile prairie and other grasslands to grow wheat for export. Some windbreaks of trees that were planted during the Dust Bowl years of the 1930s were cut down to provide more land for wheat. The United States has already lost about one-third of the topsoil of its productive farmland.[31] But a study published by the National Academy of Sciences in the mid-1980s concluded that soil erosion in the United States was unlikely to harm crop productivity in the twenty-first century. The study found that about 75 percent of the nation's cropland was eroding at low enough rates so that the lands could remain productive indefinitely.[32] A law passed by the US Congress in 1985 could continue to improve the situation, as it aims to take millions of acres of the most highly erodible land out of production by paying farmers to grow erosion-resistant grass and trees on the land. By the end of the 1980s about 30 million acres had been placed under this program.[33]

A study sponsored by the United Nations Environment Programme shows that over a 45-year period human activity has led to moderate to severe damage to the land of about 10 percent of the world's vegetated area – an area about the size of China and India combined. Activities involved in the production of food (agriculture and the grazing of livestock) caused most of the damage, and most of the land deterioration occurred in Asia and Africa. [34]

The Type of Food

As a nation develops, its diet changes. The wealthier a nation becomes, the more calories and protein its citizens consume. The average citizen of a Western industrialized nation consumes many more calories and much more protein than he or she needs for good health. Much of the excess in protein comes from a large increase in meat consumption. Often the consuming of meat instead of grains in order to get protein, which is needed for human growth and development, is a very inefficient use of food.[35] For every 16 pounds of grain and soybeans fed to beef cattle in the United States, about 1 pound of meat for human consumption is obtained. About three-quarters of the food energy in an

Street vendors sell food to many urban
dwellers in Third World cities.
(Ab Abercrombie)

Asian's diet comes directly from grain (about 300 to 400 pounds a year) whereas a US citizen consumes nearly one ton of grain per year, but 80 percent of it is first fed to animals.[36] The United States consumes the highest amount of meat per person in the world, although many other Western developed nations also have high meat consumption. It is generally agreed by experts on nutrition that excessive calories and excessive meat consumption can lead to serious health problems. Barbara Ward describes the harmful features of such a diet:

> The car and the television set and the growing volume of office work may well have produced the most literally sedentary population human society has ever known. But at the same time, diets stuffed with the proteins and calories needed for a lumberman or a professional boxer have become prevalent. Everywhere, high meat consumption demands grain-fed animals. Meanwhile, what little grain we do eat through bread usually has little nutritional value and roughage, since these are removed when the flour is refined. Thus, the human bowel is deprived of the fiber it requires to function easily.

The eating of fresh vegetables – which also give necessary fiber – has fallen off by between a third and a half in the last half century. Processed, defibered products have taken their place. The results are literally apparent. In all developed nations, obesity and diet-related illnesses are now a major medical problem….Many medical experts are now agreed that with fat, sugar, cholesterol, refined grains, food additives, and the general absence of roughage, modern citizens are literally – via heart attacks and cancer – eating and drinking themselves into the grave.[37]

Tropical rain forests are being cut down to raise beef-cattle for the US fast food market – the so-called "hamburger connection." *(United Nations)*

I'd like to end this section with a short explanation of how development has affected the first food North Americans receive after birth. If you are a North American and were born before 1940, the chances are good that the first food you received was human milk from your mother's breast, whereas if you were born after 1955,[38] your first food was probably a human-made formula from a bottle (using cow's milk as the basic ingredient). A rapid decline in breast-feeding is now taking place in the Third World, in part because of urbanization, the increasing number of women in the work force, and the promotional efforts of formula manufacturers (the latter more of a factor in the past than at present). Breast-feeding is declining also because of a desire to imitate the United States, to be "modern." As the poor in the Third World see it, if the rich are bottle-feeding their babies, it must be better.

But is it? No, it is not. Nutritionists agree that human milk is the best food for babies. Breast-feeding is also the safest, cheapest, and easiest way to feed babies. Breast-feeding probably improves bonding – a special feeling of closeness – between the mother and the baby, and, as we saw in chapter 2, it can act as a natural birth control. Breast-feeding gives babies antibodies that enable them to fight off infection; this is especially important since their own immune systems are not fully developed during the first year. A 1980 study of children of poor parents in Brazil showed that bottle-fed babies were three to four times more likely to be malnourished than those who were breast-fed. Studies have also shown that in India bottle-fed babies have diarrhea three times more often than breast-fed babies, and in Egypt infant deaths are five times higher among bottle-fed babies than among breast-fed ones.[39] Many of the harmful effects of bottle-feeding in the Third World occur because of the lack of refrigeration and the lack of knowledge about the importance of sterilization. Also, formula is expensive, so poor mothers often dilute it with water, which makes the formula too weak.

There has been a return to breast-feeding in the United States, but mothers who breast-feed their children for at least three months are still in the minority. Working mothers find bottle-feeding more convenient, and the US culture is still unsettled by the sight of a woman breast-feeding in public. Because US women can no longer turn to their mothers for help or encouragement in breast-feeding (since their mothers didn't do it), a special organization, the La Leche League, has been formed by some US women to help others learn about breast-feeding and to aid them with any difficulties they experience. What we find in this case is a modern society turning away from one of the most basic human functions and then having to relearn the advantages of this bodily function and how to practice it.

The general recognition of the harmful effects that were generated by the adoption of bottle-feeding by Third World nations led the World Health Organization in 1981 to adopt, by a vote of 118 to 1 (only the United States voted "no"), a nonbinding code restricting the promotion of infant formula.[40]

The Green Revolution ● ● ● ● ● ● ● ● ● ● ● ● ● ● ● ●

The bringing of high agricultural technology to the Third World has been called the Green Revolution. The Green Revolution has two basic components: the use of new seeds, especially for wheat, rice, and corn, and the use of various "inputs," such as fertilizer, irrigation, and pesticides. The new seeds, which were developed over 20 years of cross-pollinating, are highly responsive to fertilizer. If they receive sufficient fertilizer and water, and if pests are kept under control, the seeds produce high yields. The introduction of this new agricultural technology to the Third World in the mid-1960s brought greatly increased harvests of wheat and impressive increases in rice production in a number of Asian countries. Over a six-year period, India doubled its wheat production and Pakistan did nearly as well. Significant increases of rice production occurred in the Philippines, Sri Lanka, Indonesia, and Malaysia. Mexico's wheat and corn production tripled in only two decades.[41] Not only were the harvests much larger, but multiple harvests – in some places up to three – became possible in a year because of the faster maturing of the plants.

Unfortunately, the Green Revolution has had some significant unanticipated and negative side effects. One of the worst, which is discussed more fully in chapter 6, is that it has tended to benefit the rich and large farmers in the developing countries much more than the small, poor farmers. Fertilizer, pesticides, irrigation systems, and the new seeds all cost money, and it is the larger farmers who have the wealth to purchase these or the access to credit to finance them.

Other negative aspects of the new technology have become apparent. The new highly inbred seeds are often less resistant to diseases than are some of the traditional seeds. Also, the planting of only one variety of a plant – called monoculture – creates an ideal condition for the rapid spreading of disease and for the rapid multiplying of insects that feed on that plant. (The Irish potato blight in the mid-1800s and the US corn blight of 1970 are examples of serious diseases that have attacked monocultures.) The new seeds are also less tolerant of too little or too much water; thus droughts and floods have a more harmful impact on these plants than on the traditional varieties of the grains.

Synthetic fertilizers are usually needed with the new seeds. Fertilizer use has grown dramatically around the globe since 1970, especially in Asia and particularly in China. There is now evidence that the runoff of fertilizers from farmland is a significant source of pollution in rivers and lakes. Also, there was a large increase in the use of pesticides (insecticides, herbicides, and fungicides) around the world in the 1970s and 1980s, no doubt also connected with the spreading Green Revolution. It is difficult to know how many people are being harmed by pesticides, but it is believed that the number is significant, especially in developing countries One estimate by the World Health Organization is that perhaps as many as 20,000 deaths occur annually around the world because of pesticide poisoning and 1 million people are made ill.[42] (We will look further at pesticides in chapter 5.)

The Green Revolution also often requires irrigation. The use of freshwater, much of it for irrigation, has increased steadily since the 1960s especially in the developing nations. According to the World Resources Institute, the result has been that in many countries, for example in Africa and the Middle East, "water withdrawals appear to be occurring at unsustainable rates."[43]

Certainly, without the increased production that came with the

Green Revolution many developing countries would have already lost the battle to have enough food available for their rapidly growing populations. Dr. Norman Borlaug – a US scientist who received the Nobel Peace Prize for his work in developing high-yield wheat and the person considered to be the "father" of the Green Revolution – has stated that the Green Revolution was not meant to be the final solution for the world's food problem: it was designed to give nations a breathing space of 20 or 30 years during which time they could work to bring their population growth under control. Borlaug is as disappointed as many others are, that this time has not been used by many nations to take forceful measures to rein in their exploding numbers.[44]

Governmental Food Policies

The availability of food is such a basic need that no government that I know of adopts a "hands off" policy regarding its production, price, and distribution. As was mentioned above under the section "Causes of World Hunger," many developing nations have given a relatively low priority to agricultural development and to relieving poverty in rural areas, concentrating on industrial development instead of rural development. Nearly all of the developing nations have scarce public funds, so decisions must be made about where to apply them. It should not be surprising to students of government that public funds usually go to benefit groups with political visibility and power. Political leaders want to stay in power, and it is often the traditional political and economic elites that will influence the leaders' length of stay rather than the scattered and weak – both physically and politically – small farmers and rural poor. In many developing nations the urban masses, who can riot, are much more of a threat to the leaders than the small farmers, and urban people demand plentiful and inexpensive food.

The desire to retain power, of course, is not the only reason why rural development has not been given a high priority in many Third World nations. The desire to achieve the high living standards in the West by following the route taken by the United States and other developed nations – both capitalist and communist – with their emphasis on industrialization, has been hard to resist; it has seemed like a relatively fast way to reduce poverty. US foreign aid in the 1950s and 1960s certainly encouraged developing nations along this route. We who were in the foreign aid program then recognized that this development strategy was a gamble, that maybe benefits would not trickle down to the poor; but the other alternative of trying to work directly with the millions of rural poor did not seem viable. Barbara Ward shows how dominant this strategy of emphasizing industrialization over rural development became: "So far, on average, only 20 percent of the investment of most developing nations has gone to the 70 to 80 percent of the people who are in the rural areas."[45]

How does one respond to the argument that, given limited public funds, it is impossible to give any significant aid to the millions in the rural areas where most of the hunger exists? The response is that there have been a few Asian countries – namely, Japan, South Korea, and Taiwan – that have brought significant prosperity to their rural areas by doing certain things. First, they enacted land reform measures – in Japan's case under the US occupation forces' direction after World War II – which ended absentee landlordism and exploitative tenancy arrangements. The land was basically turned over to those who farmed it. Second, cooperatives were established to help small farmers with their purchasing of needed inputs and with the marketing of their harvests. The governments also provided information and aid to the farmers through an active agricultural extension service and by supporting agricultural research. Japanese small farmers now have some of the highest yields per acre in the world, and the mechanization they have used on their farms – mainly small machines – has tended to increase rural employment, not decrease it. Double and even triple harvests per year on the same piece of land became possible, and more laborers were required to handle these harvests.

China under Mao Zedong emphasized agriculture instead of industrialization after the disastrous "Great Leap Forward" (a crash program of economic development in the late 1950s). China has achieved impressive increases in its agricultural production, but because of its rapid population growth, the increased food has mainly gone to feed the increased number of people. Hunger is certainly less of a problem in China today than it was before the communist takeover – except during the famine in the late 1950s and early 1960s – but the costs have been high. Political opponents have been dealt with harshly and significant damage to the environment came from the efforts to increase the amount of agricultural land. Forests were cut down and marginal pastureland was converted to land for crops. Even though the communist government also made efforts to protect the environment, its actions directed toward increasing agricultural production led to an increased strain on the land. Significant increases in erosion and even possible climate changes (decreased rainfall) have been reported.[46]

Another major communist government – the Soviet Union – pursued radically different policies from China. Under Stalin's long rule, the country placed industrialization first, and agriculture was used to support that industrialization. Also, the desire to remove the political opponents of the ruling communists – the prosperous small farmers known as the "kulaks" – and the desire to substitute state-owned and collective farms for privately owned farms, led to what is commonly recognized as the destruction of efficient agriculture in that country. The Soviet Union's inability to grow enough food to feed its people caused it to import large amounts of wheat from the United States and other capitalist countries.

There is space in this chapter to sketch US food policies only briefly.

The main point that should be made is that the US government is very active in this area. Up to the 1900s the government's policy was mainly to encourage farm production, but since the 1950s the policy has been directed mainly at coping with an excess of production. The basic policy has been to prop up low farm incomes by using price supports, by purchasing surpluses, and by paying the farmers to grow less food. During the 1950s and 1960s, the policy of the US (and Canadian) governments was to buy up farm surpluses, a process that led to huge public reserves. Food from this reserve often went to poor nations under the Public Law 480 program, whereby surplus food was given or sold to developing nations. World food prices were generally stable during this period since, during bad harvest years, food from the public reserve was released. Now it is no longer the policy of either the United States or Canada to encourage large public food reserves, which means that reserves can no longer act as a cushion during periods of poor harvests. More recently, the US government has encouraged and supported the export of US farm products to other nations. The United States has become the world's leading exporter of food. The government supports this because exports help correct the large trade deficits that the country often experiences. This policy, and the strong political power of the US farmer, helps one to understand why the strongly anticommunist US President in the early 1980s – Ronald Reagan – removed the embargo on selling grain to the Soviet Union which his predecessor had ordered after the USSR invaded Afghanistan in 1979.

Future Food Supplies

How much food can be grown in the world? How many can be fed? Like most of the questions raised in this book, there are no simple answers. Also, it is not hard to find experts who give very different answers to these questions. In this final section we will look at seven topics that are directly related to these important questions: the effect of climate, the amount of arable land, energy costs, efforts to increase the efficiency of agriculture, new technology, fishing and aquaculture, and, finally, expected future food production.

Climate

Experts are in general agreement that the earth is probably going to have a warmer climate in the future. It is very difficult to predict how this will affect the world's agriculture. It could make conditions worse for the growing of food in some countries and better in others. (This subject will be dicussed more fully in the section on global warming in chapter 4). The experts are also in general agreement that there will probably be more variability in the climate than there has been in the recent past. The climate over the past several decades in the United States and

Canada has been unusually good for agriculture, but such good climate cannot be taken for granted. In fact, variability is the hallmark of the earth's climate when it is examined over long periods; one sees long-term cycles of hundreds of years and shorter cycles of 15 to 20 years.[47] Some of these cycles may be related to solar activity. A greater variability of climate (higher and lower extremes of temperature and higher and lower amounts of rainfall) will probably lower agricultural production around the world because of the large amount of marginal land that is now being used for agriculture. On land such as this, common in parts of the American West, the Canadian west, and the Russian east, a slight reduction in rainfall or a slightly shorter growing season can spell the difference between a good harvest and little or no harvest.

Arable Land

About one-half of the world's land that could be cultivated, the arable land, is presently being used for agriculture. Large amounts of potential farmland exist in Latin America and Africa; the UN's Food and Agriculture Organization (FAO) estimated in the mid-1980s that only about 20 percent of the potentially arable land in these two regions was being cultivated at that time.[48] Yet the US government's *Global 2000 Report*

Much of the food in Africa is grown and prepared by women.
(World Bank)

81

projected that land being used for agriculture would increase by only 4 percent between 1970 and the year 2000.[49] Why this big difference between the potential and the expected? One important reason is that most of the good farmland in the world is already being used. Much of the remaining arable land is far from population centers and a lot of it is marginal land, which is costly to bring into production and to maintain. Large amounts of energy would be needed to develop it – to build roads to it and transport its products to market, to irrigate it, and to fertilize it.

Because of these problems, plus the social and political obstacles that must be overcome to develop such areas, it is difficult to estimate the potential for increasing the amount of farmland. But many experts agree that only a modest increase will probably be achieved in the near future. These estimates also must take into consideration the large amount of present farmland that is being lost to agriculture through urbanization, through erosion caused by the cutting down of forests and overcropping, through the spreading of desert-like conditions (desertification) because of overgrazing and farming on the edge of deserts, and through the loss of irrigated lands (salinization and waterlogging) because of poor drainage. The *Global 2000 Report* cites as an example Egypt:

> Despite efforts to open new lands to agriculture, the total area of irri-gated farmland has remained almost unchanged in the past two decades. As fast as additional areas are irrigated with water from the Aswan Dam, old producing lands on the Nile are converted to urban uses.[50]

Also relevant to any attempt to project the amount of land that will be available in the future on which food can be grown is the matter of population growth. The World Resources Institute estimates that if pre-sent propulation growth rate projections are accurate, the "world aver-age of 0.28 hectares of cropland per capita is expected to decline to 0.17 hectares by the year 2025. In Asia, cropland per capita is expected to decline to 0.09 hectares."[51]

Energy Costs

The dramatic increase in energy costs in the 1970s had a profound influ-ence on agriculture, and expected rising energy costs in the future will strongly affect food production and the cost of food. As we have seen, modern, Western agriculture is energy-intensive, and the spreading of that type of agriculture to the Third World via the Green Revolution also entailed a commitment to using large amounts of energy. Probably more than any other single factor, the cost and availability of energy will strongly affect agricultural production in the future. In the past, a dou-bling of agricultural output has required a tenfold increase in the amount of energy used.[52] Some people hope for a breakthrough in nuclear fusion research that could lead to vast amounts of electrical energy becoming available; fossil fuels could then be designated for use in agriculture. Others hope that the Third World will somehow develop an agriculture that does not depend on the high use of energy and energy-related inputs that is common in the developed countries. To do this, it would also have to reject the Western diet as the ideal to strive for.

Increased Efficiency

If the potential for increasing the amount of farmland in the world is not great and the cost of energy can be expected to rise over the long run – as we shall see in chapter 4 – what hope is there for feeding the expanding population of the world? Some see a hope in increasing production by improving farming methods on existing farmland. The FAO estimates that vast increases in farm production could be achieved in the Third World through improved farm techniques.[53] Some of this may be achieved by spreading Green Revolution techniques to new land, but the problems of doing this, as we have seen, remain formidable.

Another possibility is the adoption of "alternative" or "sustainable" agriculture. This resource-conserving and environmentally benign agriculture utilizes a number of old, proven techniques and a new understanding of natural nutrient cycles and ecological relationships According to the World Resources Institute, it includes "practices such as crop rotation, reduced tillage or no-till, mechanical/biological weed control, integration of livestock with crops, reduced use or no use of chemical fertilizers and pesticides, integrated pest management, and provision of nutrients from various organic sources (animal manures, legumes)."[54] According to some experts, less than 5 percent of US farmers were following alternative agriculture practices in the early 1990s.[55] Whether alternative farming or some other technique is adopted – such as used on Japanese farms, where relatively small plots of land are intensely worked – an increase in efficiency on Third World farms could increase production.

New Technology

Biotechnology has been called a technology that will transform modern agriculture. Genetic engineering, the transferring of desirable genes, or traits, from one organism to another, is the best known part of this technology. New animals and plants are being created today with this technology. Plant and animal species have changed naturally throughout the evolution of life on this planet and human beings have, for thousands of years, influenced that evolution by encouraging the growth of those plants and animals having traits that benefit humans. But now, as one scientist has stated, "we can do all at once what evolution has taken millions of years to do."[56]

Biotechnology is still controversial. Its defenders point out that food crops can be developed that are resistant to insects and viruses, thus reducing the need for pesticides. On the other hand, plants can be developed that can tolerate herbicides thus allowing herbicides, which would normally harm the plant, to be used to control the weeds threatening the plant. Fruits can be developed that are resistant to spoilage. A tomato has been developed in the United States that has a natural resistance to becoming overripe, which means the tomato does not have to be picked while it is still green and relatively tasteless. Plants that are more nutritious are being developed as are plants that can grow under harsh conditions – for example, during droughts, or in salty soils, or in temperature extremes of heat and cold. With this technology animals, such as pigs, can be developed to have more lean meat, and dairy cows can be developed to produce more milk.

The critics of this new technology claim that there is a possibility that genetic engineering will alter organisims in detrimental ways that will not be fully known for years. Herbicide-resistant crops might pollinate closely related plants that are now weeds thus creating a new weed that is also resistant to herbicides. Since most of the research today in biotechnology is being performed by private corporations that see it as a way to increase their profits, it is not surprising that most of the present genetic engineering concerns crops and animals that can be profitably sold in the rich nations, not in the poor nations. The critics point to several large corporations that produce herbicides and other farm chemicals as being leaders in efforts to develop herbicide-resistant crops. Instead of encouraging the development of less reliance on chemicals in the growing of foods, this research will increase such reliance.

Like many technologies, biotechnology seems to have a positive and a negative potential. It is impossible to predict at this point which potential will dominate. It could lead to major advances in agriculture in the poorer nations. Some universities, such as the University of Ghent in Belgium, and private foundations, such as the Rockefeller Foundation in the United States, are supporting research in biotechnology that is directed toward that purpose.[57]

Fishing and Aquaculture

Not too long ago many people hoped that the world food problem would be solved by harvesting fish from the oceans, but it is now generally recognized that, as one marine biologist has put it, most of the ocean is a biological desert.[58] Nearly all the fish in the world are harvested in coastal waters and in a relatively few places further from land where there is a strong up-welling of water that brings nutrients to the surface.

Except for one period around 1970, the harvesting of fish in the world increased steadily from 1950 to the late 1980s. But a 1989 UN Food and Agriculture (FAO) report indicated that a number of the most commercially desirable fish species have started to decline because of overfishing. The FAO concluded that a more intensive effort to catch these fish would probably not lead to continued increases in the catch and in fact could cause an eventual decline in the global fish catch because of over-exploitation.[59]

One type of fishing that does hold promise for an increase in catch is aquaculture, the farming of fish inland and in coastal waters. Although aquaculture in the early 1990s was providing only a little more than 10 percent of the total fish catch in the world, the use of this technique has been growing steadily. According to the World Resources Institute, the amount of fish produced by aquaculture is expected to double by the end of the century.[60]

Nearly all the trout and catfish consumed in the United States, two common items in restaurants, are now grown on fish farms, as are nearly half of the oysters. Norway has a very large salmon breeding industry. Although aquaculture was developed in China several thousand years ago and is used extensively in some developing nations, it is now becoming more popular in the developed nations because people there – partly for health reasons (because fish are low in fat) – are consuming more fish and demanding that the fish they buy come from nonpolluted waters. Genetic engineering is also being used to create new species of fish. Here is the way one newspaper described the new techniques being used in aquaculture in the United States:

> Scientists are growing fish twice as fast as they grow naturally, cutting their feed requirement by nearly half, and raising them on a diet of groundchicken feathers and soybeans. Fish are now vaccinated against disease, sterilized so that their energy is spent growing not reproducing, and given hormones to turn females into males and males into females, changes that can be used to improve growth, taste and control of selective breeding.[61]

Future Food Production

Will the world be able to produce enough food for its rapidly expanding population? This is a hard question to answer. Many experts failed to predict the progress that has been made in food production in the last few decades, so it would be easy to discount the warnings by some of them now. Yet some disturbing signs exist. A statement by the World Resources Institute seeks to achieve a balance between opposing indicators:

> Prospects for global food and agriculture are at once promising and troubling. On the one hand, global food production has increased since 1970 and has generally been able to meet the demands of a growing world population.... .
>
> It is unclear whether production increases can continue indefinitely. Some factors augur well for global production – for example, improvements in the emerging market economies of Central Europe and possibly a multilateral agreement to liberalize agricultural trade.

In the longer term, improvements in the Soviet Union's farm economy are certainly possible. Better control of diseases (human and animal) could also open up large areas of potentially productive farming and grazing land in Africa.

On the other hand, most agricultural production in the world uses farming practices that are environmentally unsustainable. New efforts are underway in the industrialized countries to encourage more sustainable practices, but these efforts are as yet quite modest...

In developing countries, however, population growth and poverty subvert efforts to introduce sustainable practices and encourage agriculture to expand in ways detrimental to the environment. Population growth causes marginal land to be cultivated and contributes to environmental problems such as soil erosion and deforestation. . . it is far from certain that farmers will be able to adopt sustainable practices and still grow enough food to feed a projected world population of 10 billion or more people in the next century.[62]

Conclusions

One of the most fundamental problems many less developed nations face is how to end hunger in their lands. The rapid growth of their populations and the past neglect of agricultural development have resulted in increased suffering in rural areas. Advances in technology have helped to keep the overall production of food in many poor countries ahead of their increased needs, but widespread poverty in the rural districts as well as some in urban areas has meant that many people cannot afford to purchase the food that is available in the market. An emphasis on agricultural development and on increasing employment in both rural and urban areas is needed in order to provide increased income to larger numbers of the poor.

The developed nations face major food problems also. Here the problems are quite different from those faced by the developing nations. The rich nations need to learn how to produce healthful food and to retain a prosperous agricultural sector.

There are indications that among some people in the richer nations a new concern does exist with the types of food people eat. Whether this desire for more healthful foods and the awareness of the connection between food and health will spread from a minority to the majority of the people is not yet clear. It is clear, though, that in economic systems where consumers can freely exercise their preferences, the potential exists for important changes to occur fairly rapidly. For example, in the United States the relatively recent awareness of the connection between fatty foods and heart attacks has led to the production of a wide variety of low fat foods.

The picture regarding the health of the farm economy in some developed countries does not look bright. The United States has not yet learned how to maintain a sustainable, prosperous agricultural sector. Its productive capabilities are impressive, but as this chapter has pointed out, its high dependency on uncertain and potentially very costly energy supplies and its tendency to undermine the land upon which it rests, makes its future uncertain.

NOTES

1 Frances Moore Lappé and Joseph Collins, *Food First: Beyond the Myth of Scarcity*, rev. edn (New York: Ballantine Books, 1978), p. 122; and Raymond Hopkins, Robert Paarlberg, Mitchel Wallerstein, *Food in the Global Arena* (New York: Holt, Rinehart and Winston, 1982), p. 2.

2 World Resources Institute, *World Resources 1992-93* (New York: Oxford University Press, 1992), pp. 94–5.

3 Ibid., p. 96.

4 "Hunger" and "undernourishment" refer to the consumption of insufficient calories, whereas "malnutrition" refers to the lack of some necessary nutrients, usually protein. For the sake of simplicity, I am equating hunger with undernourishment and malnutrition.

5 World Resources Institute, *World Resources 1992-93*, p. 94.

6 Robert Conquest, *The Harvest of Sorrow: Soviet Collectivization and the Terror-Famine* (New York: Oxford University Press, 1986).

7 John W. Mellor and Sarah Gavian, "Famine: Causes, Prevention, and Relief," *Science*, 235 (January 1987), pp. 539, 541.

8 Hunger Project, *A Shift in the Wind*, no.12, p. 9.

9 Eric P. Eckholm, *Down to Earth: Environment and Human Needs* (New York: W. W. Norton, 1982), p. 15.

10 Roy L. Prosterman, *The Decline in Hunger-Related Deaths*, Hunger Project Papers, no. 1 (San Francisco: Hunger Project, 1984), p. ii.

11 John R. Tarrant, *Food Policies* (New York: Wiley, 1980), p. 12.

12 *New York Times*, late city edn (June 7, 1983), p. 1, and Mellor and Gavian, "Famine: Causes, Prevention, and Relief," p. 539.

13 For a fuller discussion of the causes of the African famines see Carl K. Eicher, "Facing Up to Africa's Food Crisis," *Foreign Affairs*, 61 (Fall 1982), pp. 151–74; a series of articles on Africa in the *Bulletin of the Atomic Scientists*, 41, (September 1985), pp. 21–52; and Michael H. Glantz, "Drought in Africa," *Scientific American,* 256 (June 1987), pp. 34-40.

14 Richard Critchfield, *Villages* (Garden City, NY, Anchor Press/Doubleday, 1981), p. 71.

15 Lappé and Collins, *Food First*, pp. 17-23; *New York Times*, late city edn (November 24, 1981), p. 1.

16 Presidential Commission on World Hunger, *Overcoming World Hunger: The Challenge Ahead* (Washington: Government Printing Office, 1980), p. 43.

17 Ibid., p. 45.

18 Paul R. Ehrlich, Anne H. Ehrlich, John P. Holdren, *Ecoscience: Population, Resources, Environment* (San Francisco: W. H. Freeman, 1977), p. 313.

19 Presidential Commission on World Hunger, *Overcoming World Hunger*, p. 16.

20 Ehrlich et al., *Ecoscience*, p. 303.

21 As quoted in Hunger Project, *A Shift in the Wind*, no. 15, p. 4.

22 *Christian Science Monitor* (April 1, 1982), p. 4. The productivity of US farms continued to increase throughout the 1980s and early 1990s. See Steven Holmes, "Farm Count at Lowest Point Since 1850: Just 1.9 Million," *New York Times*, national edn (November 10, 1994), p. A8.

23 Hopkins et al., *Food in the Global Arena*, p. 102.

24 William Ophuls, *Ecology and the Politics of Scarcity* (San Francisco: W. H. Freeman, 1977), pp. 42–3. The energy expended in modern agriculture is mainly nonhuman energy, of course, and most people consider that to be one of modern agriculture's most attractive features.

25 Peter Farb and George Armelagos, *Consuming Passions: The Anthropology of Eating* (Boston: Houghton Mifflin, 1980), p. 69.

26 Ibid., pp. 69–70.

27 Barbara Ward, *Progress for a Small Planet* (New York: W. W. Norton, 1979), p. 92.

28 *Statistical Abstract of the United States*, (Washington: US Bureau of the Census, 1970 and 1992), p. 582 (1970), p. 642 (1992).

29 Ward, *Progress for a Small Planet*, p. 93.

30 *Christian Science Monitor* (February 3, 1981), p. 3; (April 1, 1982), p. 4.

31 *Christian Science Monitor* (April 1, 1982), p. 4.

32 *New York Times* late city edn (May 16, 1986), p. A10.

33 Lester R. Brown et al., *State of the World* 1990 (New York: W. W. Norton, 1990), p. 65.

34 World Resources Institute, *World Resources 1992–93* , pp. 111–16.

35 The consumption of meat (and, also, milk from cows and goats) *can* make nutritional sense. Cows and sheep, for example, can consume grasses, which people are unable to digest, in places where the climate or the condition of the land makes the growing of crops impossible.

36 Ehrlich et al., *Ecoscience*, p. 315.

37 Ward, *Progress for a Small Planet*, pp. 93–4.

38 Bottle-feeding became more common in the United States than breast-feeding sometime between 1940 and 1955. A lack of good data makes it difficult to pin this down any further.

39 *New York Times*, national edn (December 17, 1982), p. 8.

40 For a discussion of the controversy over the use of infant formula see Stephen Solomon, "The Controversy Over Infant Formula," *New York Times Magazine* (December 6, 1981), p. 100.

41 Lester R. Brown, *The Twenty-ninth Day* (New York: W. W. Norton, 1978), pp. 146–7; Peter Steinhart, "The Second Green Revolution," *New York Times Magazine* (October 25, 1981), p. 48.

42 World Health Organization (WHO), *Public Health Impact of Pesticides Used in Agriculture* (Geneva, Switzerland: WHO, 1990), p. 86.

43 World Resources Institute, *World Resources 1992–93*, p. 97.

44 Population Action Council, *Popline*, 4 (August 1982), p. 2.

45 Ward, *Progress for a Small Planet*, p. 178.

46 Vaclav Smil, "Ecological Mismanagement in China," *Bulletin of the Atomic Scientists*, 38(October 1982), pp. 18–23; *New York Times*, late city edn (April 7, 1980), p. A12.

47 Tarrant, *Food Policies*, pp. 43, 279.

48 Ibid., pp. 35–6; *New York Times*, late city edn (August 17, 1981), p. D10.

49 Council on Environmental Quality and the Department of State, *The Global 2000 Report to the President: Entering the Twenty-First Century*, vol. 1 (Washington: Government Printing Office, 1980; New York: Penguin Books, 1982), p. 16.

50 Ibid., pp. 33, 35.

51 World Resources Institute, *World Resources 1992–93*, p. 96.

52 Ophuls, *Ecology and the Politics of Scarcity*, p. 54.

53 *New York Times*, late city edn (August 17, 1981), p. D10.

54 World Resources Institute, *World Resources 1992–93*, p. 100.

55 John P. Reganold, Robert I. Papendick, and James F. Parr, "Sustainable Agriculture," *Scientific American*, 262 (June 1990), p. 112. A fuller description of alternative agriculture is contained in National Resource Council, *Alternative Agriculture* (Washington: National Academy Press, 1989).

56 Harold Schmeck, "Gene-Altered Animals Enter a Commercial Era," *New York Times*, national edn (December 27, 1988), p. 17.

57 Additional arguments supporting biotechnology can be found in Charles S. Gasser and Robert T. Fraley, "Transgenic Crops," *Scientific American* 266 (June 1992), pp. 62–9. Additional criticisms can be found in Pamela Weintraub, "The Coming of the High-Tech Harvest," *Audubon* 94 (July–August 1992), pp. 92–103.

58 Ehrlich et al., *Ecoscience*, p. 353.

59 World Resources Institute, *World Resources 1992–93*, pp. 178–9.

60 Ibid., p. 180.

61 William Greer, "Public Taste and U.S. Aid Spur Fish Farming," *New York Times*, national edn (October 29, 1986), p.1. Not everyone is happy with this new technology. For some of the criticisms, see Stephen Cline, "Down on the Fish Farm," *Sierra,* 74 (March/April 1989), pp. 30–8; and Keith Schneider, "Puget Sound Fish Farms Challenged," *New York Times*, national edn (July 8, 1989), p. 6.

62 World Resources Institute, *World Resources 1992–93*, p. 94.

Further Readings

Calestous, Juma, *The Gene Hunters: Biotechnology and the Scramble for Seeds*. (Princeton: Princeton University Press, 1989). This text simplifies complicated material and explores the cutting edge of genetic research and biotechnology in agriculture.

Crawford, Michael, *The Driving Force: Food Evolution and the Future*. (New York: Harper and Row, 1989). Crawford takes a look at the possibilities for food production in the future.

Golkin, Arlene T., *Famine: A Heritage of Hunger* (Claremont, Calif.: Regina Books, 1987). Visual elements combine with text to examine the root causes of famine and to explore the devastating effect on famine's favorite victim – the world's poor.

Guttinger, J. Price, and Joanne Leslie (eds), *Food Policy: Integrating Supply, Distribution and Consumption* (Baltimore: Johns Hopkins University Press, 1987). A comprehensive survey that simplifies and explores the economic theories and problems associated with world trade, food, and famine.

Harrison, Paul, *The Greening of Africa* (New York: Penguin Books, 1987). An introduction to the vast array of problems facing the continent of Africa; details the roots of those problems and the difficulty in finding solutions.

Hunger Project, *Ending Hunger: An Idea Whose Time Has Come* (New York: Praeger, 1985). A visual exploration and discussion of hunger, including population, education, economics, and government policy.

Jensen, Bernard, *Empty Harvest: Understanding The Link Between Our Food, Our Immunity, and Our Planet* (New York: Avery Publishing Group, 1990). An interesting approach to the topic of food; discusses the issue from the perspective of world health, disease, and future population.

Kloppenburg, Jack Ralph, *First the Seed: The Political Economy of Plant Biotechnology 1492–2000* (New York: Cambridge University Press, 1988). Explores the connection between seed development and political and economic growth from the age of Columbus to the beginning of the next century.

Lipton, Michael, with Richard Longhurst, *New Seeds and Poor People* (Baltimore: Johns Hopkins University Press, 1988). Lipton looks at the development of new seeds and new agricultural techniques that often remain beyond the reach of the poor people who might receive the greatest benefit from those advances.

Rosenblum, Mort and Doug Williamson. *Squandering Eden: Africa at the Edge* (San Diego, Calif.: Harcourt Brace Jovanovich, 1987). This intensely visual book offers insight into the unique problems facing Africa and the many mouths that must be fed there.

Talbot, Ross R., *The Four World Food Agencies in Rome* (Ames: Iowa University Press, 1990). Talbot analyzes the procedures, goals, and effectiveness of the major international agencies concerned with agriculture and argues for greater US involvement in their efforts.

4

ENERGY

> *A human being, a skyscraper, an automobile, and a blade of grass all represent energy that has been transformed from one state to another.*
>
> Jeremy Rifkin, *Entropy: A New World View* (1980)

The Energy Crisis ●

Are we running out of energy? Of course not. Everything is made out of energy, and, as college students learn when they study the laws of thermodynamics in their introductory physics courses, energy cannot be destroyed. These laws also state that energy cannot be created: all we can do is to transform it from one state to another. And when energy is transformed – in other words, when it is used for some work – the energy is changed from a more useful to a less useful form. All types of energy eventually end up as low grade heat. A "law" in the physical sciences means that there are no exceptions to it, and there are none to the laws of thermodynamics.[1]

So if everything is energy and energy cannot be destroyed, why is there an energy crisis? The crisis has come because of the other laws, the laws that tell us that energy cannot be created, and that, once used, it is transformed into a less usable form. At present, the industrialized world relies on a very versatile, although polluting, fuel – oil. Oil is being consumed at prodigious rates, its supply is limited, and its price has fluctuated greatly, increasing dramatically from the mid-1970s to the early 1980s and then falling in the mid-1980s. The developed nations are facing an energy crisis because the era of cheap, and supposedly clean, energy from reliable sources is over. Table 4.1 shows this fact as well as any set of figures can, as it focuses on the changes in the price of gaso-

Table 4.1 US Gasoline prices, 1950–90

	Retail price per gallon of regular gas
1950	$0.27
1960	$0.31
1970	$0.36
1980	$1.21
1990	$1.16

Sources: *Dollars and Sense*, July-August 1980: presented in Kenneth Dolbeare, *American Public Policy* (New York: McGraw-Hill, 1982), p. 113; and *The World Almanac and Book of Facts* 1993 (New York: Pharos Books, 1993), p. 173.

line in the United States from 1950 to 1990. The table also helps one to understand another important feature of the energy crisis, especially as it has affected the United States. The period of cheap gasoline was a relatively long one, and people in the United States got used to having inexpensive petroleum products. Unprecedented economic growth and material prosperity took place in the United States during the 1950s and 1960s, and this was made possible, in part, by cheap energy. Individual lifestyles and modes of industrial production were based on plentiful, inexpensive energy, and when oil prices skyrocketed in the 1970s, the shock to the US economy, and to the economies of many other countries, was profound.

The first oil shock took place in 1973-4. The 1973 Arab-Israeli war led a number of Arab oil-producing countries to stop shipping oil to the United States and other countries allied with Israel. American motorists lined up at gas stations, vying for limited supplies. The Organization of Petroleum Exporting Countries (OPEC), of which most oil-exporting nations are members, seized the opportunity to raise oil prices significantly: they quadrupled.

The second oil shock came in 1979-80. The event that prompted this shock was the Iranian Revolution and the ousting of the Shah as the head of the Iranian government. Iranian oil shipments to the United States stopped, but the real shock came when OPEC doubled its prices. Many North Americans had refused to believe there was a real energy crisis after the first oil shock and had returned to their normal high consumption of petroleum products after the Arab embargo was lifted; but the second oil shock convinced most people that there was indeed an energy crisis. While many had blamed either the US oil companies or the US government for creating the first oil crisis, the second shock clearly demonstrated that something had fundamentally changed in the world. What became apparent to many now was that the United States, and most other developed nations, were dependent on one section of the world for a significant part of their energy, and that they could no longer control events in that part of the world.

The third oil shock came in 1990-91. Iraq invaded Kuwait and threatened Saudi Arabia. In order to prevent Iraq from becoming the dominant power in the Middle East and having significant influence on the production and pricing of oil from that region, the United States led a coalition of forces in forcing Iraq out of Kuwait. The war, which lasted just six weeks, involved a half million US soldiers and troops from other nations. A huge, sustained air attack on Iraqi forces in Kuwait and Iraq and on military facilities in Iraq (including poison gas and nuclear weapons plants) preceded the ground attack. The allied forces had few casualties, but the retreating Iraqi forces, which suffered large casualties, sabotaged more than 700 oil wells in Kuwait, setting about 600 on fire.

The United States persuaded other Western nations, including Japan, to contribute about $50 billion to help pay for the war. The United States spent about $10 billion for short-term costs. The war and its subsequent damage to their lands and economies cost all the Arab states an estimated $620 billion.[2] The price of oil increased dramatically right after the Iraqi invasion of Kuwait, but by the end of the war the price had dropped back to the prewar level. That price did not reflect the real cost of oil, which should have included the cost of the war. (It has been estimated that by the mid-1980s the United States was spending seven times as much keeping the shipping lanes open to the Middle East oil fields as it did for the oil itself,[3] and in the early 1990s it was spending an estimated $50 billion annually to keep military forces ready to enter a conflict in the Persian Gulf area.[4])

The Middle East, where much of the oil imported into the United States and Western Europe comes from, is a highly unstable area. It is torn by regional conflicts (the Arabs against Israel, Iran against Iraq, Syria against Iraq, Egypt against Libya); by religious conflicts (Moslem against Jew, Christian against Moslem, Shi'ite Moslem against Sunni Moslem, fundamentalist Moslem against secular governments); by social and ideological conflicts (traditionalists against radicals); and, in the past, by East-West competition (the United States against the Soviet Union). A large amount of the oil involved in international trade is carried on ships that must pass through a single strait in the Persian Gulf –the Strait of Hormuz.

The United States is the largest buyer of oil in the world, with much of it coming from a single country, Saudi Arabia. But many Western

European countries are even more dependent on imported oil than is the United States, as is Japan, the industrialized country most dependent on imported oil, producing virtually no oil itself and having few other domestic sources of energy.

The large increases in the price of oil by OPEC in the 1970s led to a massive transfer of wealth from the developed nations to part of the Third World. In the words of one commentator, "It may represent the quickest massive transfer of wealth among societies since the Spanish Conquistadores seized the Incan gold stores some four centuries ago."[5] Higher oil prices led to low economic growth, higher inflation, big trade deficits, and increased unemployment in the United States and other developed nations. Although developing nations use much less oil than do the developed nations, the cost of their imported oil also went up and caused some of them to acquire huge debts to pay for the oil they needed. Daniel Yergin, the coeditor of an important report on energy by the Harvard Business School, assessed the potential consequences of the oil shocks in the following terms:

> The unhappy set of economic circumstances set in motion by the oil shocks contains the potential for far-reaching crises. In the industrial nations, high inflation, low growth, and high unemployment can erode the national consensus and undermine the stability and legitimacy of the political system. In the developing world, zero growth leads to misery and upheavals. Protectionism and accumulation of debt threaten the international trade and payments system. And, of course, there is the tinder of international politics, particularly involving the Middle East, where political and social upheavals can cause major oil disruptions and where fears about and threats to energy supplies can lead to war.[6]

In the mid-1980s, the world experienced an oil glut and dramatically falling oil prices caused by significantly lower demand for oil by the industrialized nations and by conflict within OPEC over production goals (partly caused by a long and bitter war between two of its members – Iran and Iraq). A serious recession in the early 1980s in both the United States and in Western Europe – the worst in the United States since the Great Depression of the 1930s – led to a reduced demand for oil, a demand that was reduced further by conservation measures, and by the switching to natural gas by some oil consumers. Stimulated by the high oil prices of the 1970s, non-OPEC oil producers increased their production, contributing to a plentiful supply of oil. The falling oil prices in the mid-1980s caused a number of Western oil companies to close some of their oil wells that were costly to operate and to reduce their budgets for the exploration of new oil.

While many expected the Persian Gulf War in the early 1990s to lead to a big increase in the price of oil, it did not occur. Saudi Arabia increased its production, and demand by the West was relatively low because of a continuing recession and a warm winter. In the short term the price of oil may remain low, as Saudi Arabia remains indebted to the United States for protecting it from Iraq and is aware that any large increase in price would encourage the West to search for alternative sources of energy. But the longer-term prospects for the price of oil remain uncertain. In the early 1990s governments hostile to the West remained in power in Iraq and Iran, Iran was involved in a buildup of its military power, the long-term stability of the Saudi government was uncertain, and the US reliance on imported oil was growing.

I have focused in this section on the oil crisis, which has affected mainly the industrialized nations. But that is not the full story of the energy crisis that the world faces. About one-half of the world's population uses no fossil fuels at all, relying mainly on wood, charcoal, cow dung, and crop residues for cooking fuel and for heat.[7] The shortage of firewood in the Third World is increasing as population growth has caused consumption of wood to exceed the growth of new supplies in many areas. Forests are being cut down and are not being replanted. The dependency of the poor in the Third World on wood is in some ways like the dependency of the rich on oil. Both dependencies can be dangerous and will require forceful public and private measures to be reduced.

Shortage of wood is a part of the energy crisis, since many urban dwellers in developing nations rely on wood as their major source of fuel. (*Ab Abercrombie*)

Responses by Governments to the Energy Crisis ● ● ● ●

Let us look at a few key countries and regions to see how their governments have responded to the energy crisis.

The United States

The US response to the energy crisis has been rather feeble. No coherent policy for dealing with the crisis has been adopted although a number of laws dealing with the crisis have been passed. President Richard Nixon called for "Project Independence" to make the United States self-sufficient in energy by 1980, and President Jimmy Carter believed that the energy crisis should be considered the "moral equivalent of war." But in fact, the US response did not seriously reduce the country's dependency on oil. Why is the United States having difficulty enacting an effective policy to deal with the crisis? Part of the reason is that the inertia of an oil-intensive society is hard to overcome. The nation is used to abundance, in energy as well as in material goods, and the creation of a new outlook and new values is not easy.

The cost of oil in the United States, in one sense, remains very low. When inflation is taken into account, the price of gasoline in the early 1990s was about equal to the price of gasoline in the mid-1940s.[8] What this means is that the "real," or true, cost of oil was not indicated by its price and thus consumers in the United States felt no urgency in demanding – or the government in producing – an energy policy that would break the dominance of oil in their society. As shown by some energy analysts, the real cost of oil would have to reflect not only the military costs necessary to secure its supply, but also the costs of the environmental degradation it causes – such as the spilling of 11 million gallons of oil in the coastal waters of Alaska by the tanker *Exxon Valdez* in 1989. The real cost of oil would have to reflect also the increased health care costs that come with its use and the subsidies by the government to the oil industry.[9] Gasoline sales taxes can be used to cover some of these hidden costs – which are borne by the whole society – but as shown by figure 4.1, the tax on gasoline in the United States has remained much lower than that in other major industrialized nations.

Another reason for the failure to produce an effective energy policy, is that important groups in US society have conflicting goals relating to energy. The economist Lester Thurow gives the following answer to the question of why the United States has been unable so far to take forceful action to deal with the crisis:

> The lack of action does not spring from a lack of solutions, but from the fact that each solution would cause a large, real income decline for some segment of the population. Everyone is in favor of energy independence in the abstract, but each path to energy independence

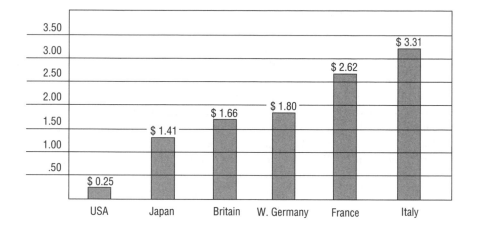

Figure 4.1 Gasoline tax per gallon (1990).
Sources: Energy and Resources Branch, UN Secretariat, as presented in *New York Times*, national edn (September 24, 1990), p. C5.

is vigorously opposed by some significant group that would suffer large income declines if this particular solution were chosen. In the process, all solutions are vetoed and we remain paralyzed. The status quo is painful, but we cannot move.[10]

After the Soviet Union invaded Afghanistan, the US goverment in 1980 announced, in what became known as the Carter Doctrine, that it would use military force, if necessary, to keep Middle Eastern oil supplies in friendly hands. The United States created a rapid deployment military force designed to enable US forces to fight in the Middle East on short notice. This military force was called into action when a major oil supplier of the West – Kuwait – was invaded. The invader, of course, was not the Soviet Union, which was disintegrating, but neighboring Iraq.

The US government's support for nuclear power has been strong. More governmental funds for research and development have been spent on this energy source than on any other. Funds designated for the promotion of solar energy and other renewable forms of energy increased under the Carter administration in the late 1970s, but the Reagan administration in the 1980s drastically reduced governmental support for such development. The Reagan administration cut governmental funds for renewable energy research by nearly 90 percent and removed the solar panels from the White House roof. Some laws designed to promote the conservation of energy were passed, such as the 55-mile-per-hour speed limit, tax credits for home insulation, and higher fuel efficiency standards for automobiles, but most of these laws lapsed in the 1980s.

For many years the government controlled the price of oil and natural gas in the United States, partly to protect the profits of US producers. In order to promote the exploration of domestic sources of oil and natural gas and to encourage conservation by higher prices, all controls on oil prices were removed in the early 1980s and a phased reduction of price controls on natural gas was enacted. In 1980 Congress approved a multibillion dollar program to develop synthetic fuels, mainly through the conversion of large deposits of coal and shale into oil and gas, but in 1985 Congress terminated the program because of falling oil prices and high construction costs. Policies were adopted to encourage electric utilities to shift to using coal instead of oil or natural gas. Finally, a strategic petroleum reserve was established in the mid-1970s and by the early 1990s the United States had a two-month supply of oil stored in salt caverns along the Gulf coast.

In the early 1990s the United States tried to create an energy strategy in response to the Persian Gulf War. The law that finally passed Congress and was approved by President Bush again failed to reduce the country's dependence on imported oil, which by that time had risen to be about one-half of the country's requirements. The law did provide for the streamlining of the licensing of new nuclear power plants, and it did require higher efficiency standards for many household appliances, lights, and motors. Potentially more significant than this bill for changing the use of energy in the United States was the Clean Air Act of 1990, which was enacted to encourage the use of methanol, natural gas, and electricity to run cars. California, in order to reduce its serious air pollution, has issued regulations that will require the sale of thousands of vehicles that will run on alternative fuels. Several large urban states in the Northeast have adopted similiar regulations.

In summary, we can say that, up through the early 1990s, US policy emphasized securing – even through military action – and producing more fossil fuels; continuing a strong commitment to nuclear power; promoting, deemphasizing and then starting to promote again renewable energy; and promoting energy conservation, first through governmental programs and later by relying on higher prices to decrease use of oil and natural gas.

Western Europe

As mentioned above, most Western European countries are more dependent on imported oil than is the United States. Traditionally, European governments have let the prices for imported fuel go up as the world market determined and have tried to encourage energy conservation through the use of high taxes. France has emphasized nuclear power as its response to the energy crisis, and by the early 1990s it was producing about 75 percent of its electricity from that source – a world record. West Germany has created a national oil company to explore for oil around the world, is continuing to construct nuclear power plants, although there is some public opposition to nuclear power; it also is purchasing natural gas from the former Soviet Union, and is expanding coal production. The discovery of oil and natural gas under the North Sea has aided mainly Norway and Britain. This large deposit may allow Britain to become self-sufficient in oil and natural gas, at least for a while. Coal remains a major fuel in Central Europe and the former Soviet Union, where it has produced severe air pollution.

Japan

Japan has no significant oil, natural gas, or coal deposits; as stated above, it is the most vulnerable of all industrialized countries to OPEC's actions. A consensus quickly developed in the country, after the first oil shock in 1973, that its dependency on oil must be reduced. The government encouraged conservation and the people responded. Daniel Yergin reports that "In 1973, Japan used only 57 percent as much energy for every unit of GNP as did the United States. By 1980, it used only 43 percent as much."[11]

It is interesting to note some of the differences between Japanese and US societies that have undoubtedly affected their different responses to the energy crisis. Because of their history and their limited land and resources, the Japanese have always assumed scarcity and insecurity of resources such as fuel, whereas the Americans have been accustomed to abundance and have assumed it will continue. Japanese industries have been traditionally more willing than their US counterparts to make long-term investments, the American companies often being more concerned with making short-term profits. The Japanese know that their goods must compete well in international trade if they are to maintain their high living standards. Japan is used to change and adaptation. The consensus that developed in Japan after 1973 emphasized a shift from consumption to restraint. It included a belief that the economy had to shift to "knowledge-intensive" industries that use relatively little energy, and that energy efficiency was the key element in the adjustment the country needed to make to this new situation.[12]

Japan made significant progress in the period between the oil shocks in the 1970s and the third one in 1990–1. By 1990 the energy efficiency of the Japanese economy had improved to such an extent that the production of goods and services took only one-half the energy it took in the late 1970s.[13] The increased efficiency in the automobile and steel industries came after the government set ambitious goals for them to reach.

Another action taken by the government after the early crises was to build large oil storage facilities. By the time of the third oil crisis, Japan had nearly a five-month supply of oil in storage tanks, more than any other nation. The country also sought to diversify its sources of oil. By the time of the third shock, it had reduced its oil imports from the Middle East so that about 70 percent of its oil came from that region, down from 80 percent during the previous shocks.[14] The Japanese government also has a billion-dollar research program, called Operation Sunshine, designed to make the Japanese the leader in solar energy technology.

The Japanese government has made nuclear power one of the key parts of its plans to reduce its dependency on imported oil. In the early 1990s nuclear power provided about 25 percent of the electricity in the country, and the government had plans to increase that to about 45 percent by the early part of the next century. Japan's future plans include the construction of a number of fast-breeder reactors, to reduce its dependency on imported uranium. In the early 1990s Japan began importing plutonium from France (recycled from spent uranium fuel from Japanese power plants) for those reactors.

China

Although China's energy situation is atypical for a Third World country because of its vast reserves of coal, it does have a typical Third World problem: how to provide a growing population with enough fuel in a manner that does not seriously harm the environment. China's population is so large that its use of energy could have a significant effect on the world's environment, as the following excerpt makes clear:

> The path of industrial development in China... could have a greater effect on the atmospheric accumulation of carbon dioxide than that of any other nation. China's critical role stems from its large and growing population, its tendency toward energy-intensive processes, its poor energy efficiency, and its massive reliance on coal.[15]

China has the largest coal deposits in the world and relies mainly on coal for its energy needs in its urban areas. This extensive use of coal is creating major air pollution and other environmental problems. Most of this coal is situated in the northwestern part of the country, far from the eastern coastal provinces where much of the new economic growth is taking place. Factories in Shanghai and in other cities in this region must shut down for several days a month because of a shortage of energy.[16] It is mainly for this reason that China plans to build more nuclear power plants in the industrialized coastal areas.

China has the world's largest program to create methane gas for use as fuel in rural areas. The gas is produced by fermenting animal and human wastes in simple generators; after the gas is produced, a rich organic fertilizer remains that can safely be used on crops.[17]

Most Chinese live in villages and hundreds of millions of poor peasants continue to rely on inadequate amounts of crop residues, dried animal dung, and forest wood for their energy. According to Vaclav Smil, the "staggering dimensions" of China's rural energy shortage are seen in the fact that "about half a billion people ... [lack] enough fuel just to cook three meals a day for three to six months a year!"[18]

Energy shortages in Rural China

A Canadian geographer describes where Chinese villages get some of their fuel and the harmful effects of their actions on the environment:

Fuel comes from literally any burnable tree and forest biomass: not only branches and twigs, roots and stumps, but also bark off of the living trees, needles, leaves and grasses, and carved-out and dried pieces of sod. People carefully raking any organic debris accumulated on the floor of even small groves, peasants drastically pruning the summer growth of shrubs and trees, and children gathering tufts of grass into their back baskets are common sights in China's fuel-short countryside.

Environmental consequences of this often desperate search for fuel are predictably severe: rapid nutrient loss and erosion of slope sites stripped of trees or of the protective floor-debris cover not infrequently result in the total loss of the site for eventual revegetation.

Source: Vaclav Smil, *Energy in China's Modernization: Advances and Limitations*, (Armonk, NY: M. E. Sharpe, 1988) p. 51–2.

The Chinese use relatively little oil, given the size of China's population, and the country is able to export some of its output. For a number of years there was hope that oil production would dramatically increase in China as the country invited foreign oil companies to explore for oil reserves in its offshore waters. But by the early 1990s no major discoveries had been made, and the hope that China would be able to significantly increase its production started to fade. Possibilities still exist for major discoveries inland, but China has been reluctant to let foreign companies search there. As China continues to industrialize, its consumption of oil will increase, even though it still relies primarily on coal for its energy. If new discoveries of oil are not made, China could change from an exporter to an importer of oil by the mid- to late- 1990s.[19]

The Effect of the Energy Crisis on Third World Development Plans

The early stages of industrialization are energy-intensive. Modern transportation systems, upon which industrialization rests, utilize large amounts of energy, as does the construction industry. The huge increase in oil prices in the 1970s cast a cloud over the development plans of many developing nations. Most of these plans were based upon an assumption that reasonably cheap oil would be available, as it had been for the West, to support their industrialization. Most of the Third World countries have little or no coal or oil themselves. The development plans called for these countries to export natural commodities, nonfuel resources, and light manufactured goods; it was assumed that the earnings from these exports would be sufficient to pay for the fuel they would need to import. The success of the development plans also depended upon the countries being able to generate enough capital locally so that funds for investment in businesses would be available.

When OPEC increased fuel prices, no exceptions were made for the poorer countries; they were required to pay the same high prices for their oil imports as the rich nations had to pay. Added to that burden was the one created by the global recession that the higher oil prices had helped to create. As the recession deepened in the West, the industrialized countries cut back on their imports from the developing nations. Many of these countries borrowed heavily from commercial banks to pay for their higher oil bills and accumulated staggering debt. The World Bank estimates the foreign debt of the less developed countries in the early 1990s to be about $1.3 trillion.[20] Brazil had the largest foreign debt of all the developing nations (over $100 billion in the early 1990s) and was having serious troubles in trying to repay it.[21]

The replacing of human-powered vehicles with oil-fueled vehicles in poor and crowded countries, such as Bangladesh, does not appear to be possible. (*World Bank*)

The new situation created by high oil prices has led some experts to talk about a "Fourth World." This term refers to some of the developing nations, such as Bangladesh, which have few natural resources of their own and little ability to purchase the now expensive oil to promote industrialization. The countries in the Fourth World, the "poor poor," have little chance of ever developing along the path followed by the West with its dependency on fossil fuels. If these nations are to improve their living standards, they will have to follow a development path radically different from the one followed by the developed nations.

Many energy experts believe that there is little hope that conservation can help very much to improve the energy crisis in the Third World since there is little waste of energy there now. And many experts predict that the largest increase in demand for oil for the remaining part of this century will come from the industrializing Third World nations with high population growth, and not from the developed nations, which have low population growth and are becoming more energy-efficient.[22]

Population pressure and the high cost of oil are increasing the demand for traditional fuel in the Third World, which is mainly wood. This problem has been mentioned above and will be discussed further in chapter 5, which discusses the environment. As firewood becomes expensive or unavailable in rural areas, people switch to burning dried cow dung and crop residues, thus preventing important nutrients and organic material from returning to the soil.

The Relationship Between Energy Use and Development

A Shift in Types of Energy

One way to study the progress of the human race is to focus on the way humans have used energy to help them produce goods and services. People have constantly sought ways to lighten the physical work they must do to produce the things that they need – or feel they need – to live decently. The harnessing of fire was a crucial step in human evolution as it provided early humans with heat, enabled them to cook their foods, and helped them to protect themselves against carnivorous animals. Next came the domestication of animals. Animal power was an important supplement to human muscles, enabling people to grow food on a larger scale than ever before. Wood was an important energy source for much of human history, as it still is for a large part of the world's population. The replacment of wood by coal to make steam in Britain in the eighteenth century enabled the Industrial Revolution to begin. In the late 1800s oil, and in the early 1900s natural gas, began to replace coal since they were cleaner and more convenient to use. Oil had overtaken coal as the principal commercial energy source in the world by 1970. In the 1970s nuclear power was introduced and was producing about 20 percent of the world's electricity by the late 1980s.

Increased Use

The use of energy in the world has increased dramatically in the years since the end of World War II, a period of rapid development in the industrialized countries and one marking the beginning of industrialization in a number of Third World countries. Up to 1990, most of the increased energy use took place in the developed nations. The United States was the largest user of energy from 1950 to 1990: in 1990 the United States, with about 5 percent of the world's population, was using about one-quarter of the energy being used by all nations. The United States consumes more energy annually than does all of Europe, even though the population of the latter is double that of the former. As Figure 4.2 shows, the United States also leads the world in per capita consumption of energy. Although the energy use per person in the United States exceeded that of other developed nations in 1989, the difference between the energy use of the United States and that of the developing nations was so great as to be hard to comprehend. Per capita energy consumption for commercial activity in India in 1989 was about 700 pounds of coal equivalent of energy, while in the United States it was about 22,000 pounds. In Bangladesh it was only about 150 pounds.[23]

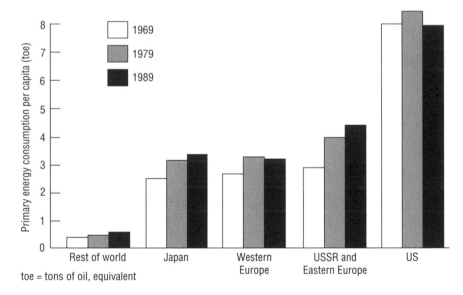

toe = tons of oil, equivalent

Figure 4.2 Energy consumption per capita.
Source: British Petroleum Company, *Statistical Review of World Energy* (London: British Petroleum Company, 1990).

People in the United States use a lot of their energy for transportation, and that means automobiles, and more recently trucks.[24] US citizens are the most mobile people in the history of the world. They are the world's modern nomads: instead of using camels or horses as traditional nomads do, they use cars. An examination of what has happened to cars in the United States shows one reason why US energy use increased dramatically in the 1950s, 1960s, and early 1970s. Let's look at what happened to one automobile model – the Ford Thunderbird. Table 4.2 shows that the length, weight, and engine size of Thunderbirds increased from 1955 to 1975. The price of gasoline was low and stable during most of the years when the model was getting bigger, heavier, and more powerful.[25] After there was a large increase in the price of gasoline, the car became smaller, lighter, and less powerful in 1980.

Table 4.2 The evolution of a US automobile: The Ford Thunderbird

Year	Length	Weight (pounds)	Engine size (displacement in cubic inches)	Leaded gasoline (dollars per gallon)
1955	14' 7"	3,300	290	0.29
1960	17' 1"	4,300	350	0.31
1965	17' 1"	4,600	390	0.31
1970	17' 8"	4,600	430	0.36
1975	18' 10"	5,100	460	0.57
1980	16' 8"	3,200	260	1.23

Source. Adapted from *New York Times*, city edn (July 21, 1980), p. A12.

We cannot fully understand the increased use of energy in the United States by focusing on what was happening to just one automobile model, of course, but such a focus does help to clarify an important part of the story. Heavier and more powerful cars use significantly more fuel than lighter and less powerful ones. The trend in US cars in the decades before the oil shocks was in the direction of more weight and greater power. Furthermore, the key role played by the automobile in the US economy is illustrated by the fact that, in the mid-1970s, six of the world's ten largest companies that were based in the United States, sold either oil or cars.[26]

Recession, Higher Prices, and a Reduction in the Demand for Oil

As we saw, after the first oil shock in 1973, most US citizens returned to their old ways of high consumption of oil and other forms of energy. The panic that came with the oil embargo and led to a high demand for smaller, more fuel-efficient autos, soon passed once the embargo was lifted and gasoline supplies were plentiful again. Even though gasoline prices were higher after the embargo, they were not high enough to discourage people from buying large cars. But when the second oil shock hit the West in 1979–80, coming as it did during a recession, many US consumers came to the conclusion that the energy crisis was real. The result was a dramatic shift to purchasing small automobiles and driving less. Many turned to buying cars imported from Japan, which were recognized as being well made and much more fuel-efficient than most US models. The rush to buy imports caused a crisis in the US automobile industry since there was no longer a large market for its big cars. The industry has always preferred selling larger cars to smaller cars because their profits are higher on the larger models. But after much agony, the near collapse of one of the large automobile companies (Chrysler), and a large financial investment, the US automobile companies shifted to producing smaller cars with good fuel-efficiency ratings. With the drastic fall of oil prices in the mid-1980s, however, many people in the United States again started to show a preference for larger and more powerful automobiles.[27]

The Decoupling of Energy Consumption and Economic Growth

Historically, there has appeared to be a one-to-one relationship in the United States between economic growth and energy growth; for example, a 10 percent increase in the amount of goods and services produced in the country was accompanied by an approximately 10 percent increase in the amount of energy consumed. But the oil shock of 1973 seems to have broken this relationship. Between 1973 and 1986 the US economy grew about 35 percent but the amount of energy used remained about constant.[28] What happened was that the United States had begun to use energy much more efficiently than it had before 1973, no doubt in response to the higher oil prices. But then in 1986 the price of oil fell dramatically and remained relatively low. Probably in large part because of that fact, the efforts to further conserve energy in the United States came to a halt. From 1986 to 1992 the amount of energy needed to produce goods and services worth a dollar remained nearly level.[29]

This partial decoupling of energy use and economic growth is not surprising once one realizes that there are a number of countries with high levels of economic prosperity that have traditionally used much less energy than does the United States. Sweden, a country with a higher living standard than the United States and higher heating requirements, uses about half the energy per person that the United States uses. Germany, another wealthy country that has had impressive economic growth, also uses about half the energy the United States uses per capita. And Japan , which even before the oil shocks used about half the energy the United States used to produce goods and services, so improved its energy efficiency after the first two oil shocks that by the early 1990s it was using half the energy to produce $1 worth of goods and services that it used in the late 1970s.[30] In 1990 the United States was using about 40 percent less oil and gas to produce $1 worth of goods and services than was used in 1973.[31]

A number of studies of the US energy situation have concluded that a more efficient use of energy can actually lead to economic growth.[32] A study prepared for the US president recommending actions the country should take to deal with the situation presented in the government's *Global 2000 Report*, describes some of the energy inefficiencies in the US economy:

> Evidence is mounting that US economic growth, as measured by Gross National Product (GNP), need not be tied to a similar energy growth rate. The most important reason is that the US economy, including much of its building and transportation stock, its industrial processes and machinery, is inefficient in its use of energy, compared both with other economies and with the technological and cost-effective options that already exist. The opportunity is enormous for improving the energy efficiency of US capital stock – in effect creating "conservation energy" – to get the same desirable end result of warmth, comfort, jobs, and mobility that fossil fuel energy provides.[33]

Energy conservation can promote economic growth because the cost of saving energy through such measures as improving the fuel efficiency of cars, improving the efficiency of industrial processes, insulating houses, etc., is lower than the cost of most energy today. Also, investments in improving the energy efficiency of US autos, homes, and factories create many new jobs and businesses throughout the country, thus spurring the growth of the economy in contrast to draining funds out of the economy by purchasing foreign oil.

Part of the reason many European countries use much less energy per person than does the United States is that they are smaller countries with populations not nearly as dispersed. One study has shown that the long distances people and goods move in the United States, in contrast with Europe and Japan, and the US preference for large, single-family homes, accounts for about 40 percent of the difference between high US energy use and lower foreign use. The other 60 percent of the difference is accounted for by the fact that the fuel economy of US auto-mobiles has historically been much poorer than that of many foreign cars and the energy consumption per unit of output of many American manufacturing firms is higher than that of the foreign companies.[34]

The United States obviously cannot do anything about its size, but there are things that can be done to improve the energy efficiency of its transportation equipment. As mentioned above, the federal government passed a law in 1975, over the strong opposition of the automobile indus-try, requiring the fuel efficiency of American automobiles to be gradually improved.[35] About 90 percent of the long-distance hauling of freight in the United States is by truck, a vehicle that uses four times as much energy to move a ton of freight as does a freight train.[36] The US government, through its vast expenditure of funds on the interstate highway system (reported to be the largest public works project in history), its much lower tax on gasoline than in Europe and Japan, and its relatively small amount of expenditures that benefit the railroads, has done much to promote the use of trucks over trains in the country. This policy could be reversed.

Global Warming

Many scientists believe that the human race is now involved in an experiment of unprecedented importance to the future of life on this planet, involving nothing less than the global climate. A change in the global climate may now be taking place, mainly because of the burning, by humans, of large amounts of fossil fuels – coal, oil, and natural gas. When these fuels are consumed, the carbon that has accumulated in them over millions of years, is released into the atmosphere as carbon dioxide (CO_2). Scientists are in general agreement that CO_2 in the earth's atmosphere has increased significantly since the Industrial Revolution: by about 25 percent between the mid-1700s to the present.[37] This increase, according to many scientists, will cause a warming of the earth's surface – called "global warming" or the "greenhouse effect" – since CO_2 in the atmosphere allows sunlight to reach the earth but traps some of the earth's heat, preventing it from radiating back into space. While CO_2 is the largest contributor to possible global warming, other gases are increasing significantly and contributing to this process. These include methane, which comes from both natural and human causes; nitrous oxide, which comes from fertilizers and other sources; and chlorofluorocarbons (CFCs), widely used in air conditioning. [38]

There is no controversy over the increase in CO_2 levels in the earth's atmosphere, but there is some controversy among scientists as to whether the increasing CO_2 and other so-called greenhouse gases will actually cause a warming of the earth. There is strong evidence that in the past CO_2 and methane in the atmosphere were closely connected in some way with the earth's temperature. A French-Soviet team of scientists in the Antarctic drilled a hole about one mile deep in the ice and withdrew a core of ice. Like the rings of a tree, the core indicated changing conditions in the past – in fact, back about 160,000 years. The scientists measured the amount of CO_2 and methane in the air bubbles in the ice and found two amazing facts. First, the amount of CO_2 in the earth's present atmosphere is higher than in any previous time during those 160,000 years, and the levels of CO_2, and methane, and the earth's temperature went up and down closely together during that period.[39] What the scientists cannot know from this close correlation is whether the CO_2 and methane caused the temperature change, or whether the temperature first changed and that caused the change in CO_2 and methane or even if a third factor or additional factors caused the CO_2, methane, and temperature to change. But the close connection between the earth's temperature and the methane and CO_2 levels is consistent with the global warming theory.

There is evidence that over the past century the temperature of the earth has increased by about one degree Fahrenheit (about half a degree Celsius).[40] And as of 1991, the ten warmest years since 1880 occurred after 1972. [41] Other evidence supporting the assertion that global warming has begun is the fact that most mountain glaciers in the world have been retreating since the late 1800s and over the same period the level of the oceans has risen by an average of 1 to 2 mm per year.[42] But scientists are not sure that these occurences are an indication that global warming has begun or just an indication of the natural change in the earth's temperature as it goes through its normal cycles. Many scientists now believe that it will not be until the beginning of the twenty-first century that evidence will be unequivocal that global warming is occurring.[43]

Numerous models of the earth's climate have been made by climatologists and nearly all of these predict a warming of the earth because of the increasing CO_2 and other greenhouse gases. The most common forecast of the models is that – based on present trends – the amount of CO_2 in the atmosphere is expected to double over preindustrial times by about the middle of the next century and that will lead to an increase of about 5° F (3° C) before the end of the next century. [44] While five degrees does not sound like very much, it would be a significant change. According to scientists of the US National Aeronautics and Space Administration, the temperature on earth "would approach the warmth of the Mesozoic, the age of dinosaurs."[45] There would be major changes in the amount of rainfall and its location, with some areas getting more rainfall than at present and some less. Parts of the world will have a better climate for growing food and some will have a worse one. Scientists are unable to predict reliably which areas would be hurt and which would gain, but there is speculation that some of the major food-growing regions of the world at present, such as in the United States, would be seriously harmed by the climatic changes. It is estimated that much of the central United States, the area where much of the nation's wheat and corn is grown, would have hot, dry conditions. India could have better climate for agriculture than it has now.[46]

Another probable effect of a warming of the earth's climate is that the level of the oceans will rise. Such a gradual rising of waters could lead to the evacuation of some coastal cities around the world. The rich countries would probably be able to build dikes to protect their cities, but poor countries such as Bangladesh could probably not afford to do so. Also much coastal lowland around the world would be threatened. These lands are heavily populated at present, especially in the developing nations. It is now predicted that the oceans will rise by about half a foot (20 centimeters), by 2030 and by about 2 feet (65 centimeters) by the end of the next century.[47]

Critics have attacked the predictions of global warming on several counts.[48] They point out that the models of the earth's climate that climatologists use today to make forecasts are deficient, especially in their understanding of how the oceans and clouds will affect the possible warming. They might either speed it up or slow it down. In fact, the critics point out that we do not understand enough about possible "positive feedbacks," those things possibly occurring if the warming takes place that will make it worse – such as a warming of the permafrost could release more methane. We also do not understand about possible "negative feedbacks," those things that could make it cooler – such as an exploding algae population in a warmer ocean could absorb more carbon dioxide.[49] The defenders of the global warming theory admit that the climate models they use to make predictions are still relatively crude, but as of the early 1990s the critics were in the minority. About 400 scientists from 25 countries contributed material to the most authoritative report on global warming in the early 1990s, and they concluded that because of the release of "greenhouse gases" by humans, an increase in the earth's temperature would occur. They also concluded that, because of various "feedbacks," it was likely that the warming would be even more than they were predicting They also warned that because of our incomplete knowledge about the processes involved in the earth's climate, it is possible we will be confronted with "surprises" in the future. [50]

The world faces a real dilemma concerning this situation. A report by the US Council on Environmental Quality, which examined many of the relevant scientific studies on the CO_2 problem as of 1980, explained the predicament:

> The carbon dioxide problem poses an extraordinarily difficult dilemma for the international community of nations. To respond in a significant way now to a threat whose scope and time of onset are still uncertain might require unnecessary commitments of resources. To postpone taking action until a substantial climate change were detected – which could be 20 years away – would entail a risk of being unable to prevent long-term climate changes that could prove serious and irreversible for centuries.[51]

There *are* policies that the United States and other nations could pursue that would alleviate this threat but would not require them to make a serious sacrifice at present. One would be to deemphasize programs to promote the increased use of coal and synthetic fuels made from coal and oil and to stress the conservation of energy and the development of renewable energy sources, such as solar energy, and nonfossil fuel energy, such as nuclear energy. In the short term, the United States and other high-energy users could switch from oil to natural gas as natural gas releases 30 percent less CO_2 per equivalent amount of energy than petroleum (compared to coal, natural gas releases about 40 percent less CO_2). Another policy would be to combat deforestation, since trees, along with other vegetation, absorb large amounts of carbon dioxide.[52] (The increasing destruction of the great tropical rain forests in Latin America is seen by some experts as representing a real threat to the global climate).

As figure 4.3 shows, the industrialized nations produce most of the CO_2 at present. Each person in countries such as the United States and Canada presently produces an average of about 20 tons of carbon dioxide each year because of their high energy consumption, while each person in the developing countries produces an average of about 3 tons.[53] But as development spreads to some of the large less developed nations – such as China and India – and as their population grows, it is expected that they will produce a relatively larger percentage of the gas, especially as China relies mainly on coal, the fossil fuel emitting the most CO_2. How the less developed nations can be encouraged to develop without increasing their CO_2 emissions is not clear, but it is obviously in the interests of the industrialized nations to help them do so. It is also in the interests of both rich and poor nations to support population control efforts in the developing world, as more people will release more carbon dioxide and other greenhouse gases.

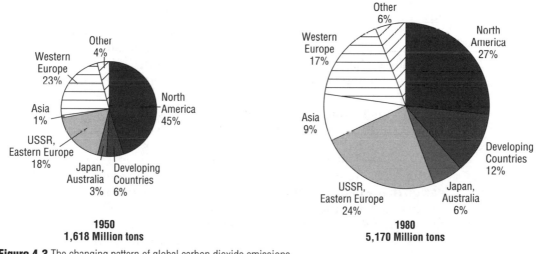

1950
1,618 Million tons

1980
5,170 Million tons

Figure 4.3 The changing pattern of global carbon dioxide emissions.

Source: Robert M. White, "The Great Climate Debate," *Scientific American* (July 1990), p. 41.

In June 1992, at the UN-sponsored environmental conference held in Rio de Janeiro, about 150 nations signed a treaty that provided for reducing the emissions of CO_2 and other greenhouse gases to 1990 levels. European nations, Japan, and about 40 small island and coastal states favored putting specific targets and timetables for reaching those targets in the treaty, but the United States opposed this and they were not included in the treaty.

The United Nations is coordinating scientific research efforts directed at the global warming problem, but, as the US government report cited above states, the world probably cannot afford to wait until the evidence is clear that human activity is causing a climate change. The view of Gus Speth, the former chairman of the US Council on Environmental Quality, is as follows:

> With our limited knowledge of its [the earth's] workings, we should not experiment with its great systems in a way that imposes unknown and potentially large risks on future generations.... People have altered the face of the planet throughout history, but the power of today's technology and our growing capacity to foresee, however uncertainly, the possible consequences of our acts puts us in a new moral position."[54]

The Energy Transition

The world is entering a period of transition from one main energy source – oil – to a new principal source or a variety of sources. This is the third energy transition the world has passed through: the first was from wood to coal, and the second from coal to oil. Many people, although not all by any means, now recognize that the industrialized world must shift from its reliance on depletable, nonrenewable fossil fuels to an energy source or sources that are renewable or, for all practical purposes, nondepletable. Many in the United States and in the other industrialized countries understand that their dependence on imported oil must end since it is no longer cheap, abundant, or secure. But what will be the new principal energy source for the industries of the developed world and for those of the developing nations? As in many periods of transition, the end to be reached is not clear. The only clear thing now is that the old state of affairs is no longer viable.

Since we are only in the beginning years of the energy transition, a period that will no doubt take many years to pass through, it is interesting to note what the first response by the largest energy consumers in the world was to the need to develop new energy sources. In the late 1980s the member countries of the International Energy Agency allocated 60 percent of their energy research and development budgets to nuclear power; 15 percent to coal, oil, and gas; about 20 percent to electric power transmission; and only about 5 percent to renewable energy sources.[55]

For the rest of this section, we shall examine some of the potentialities of the most often discussed energy sources, and in the final section we shall focus on nuclear energy, looking in some detail at the main arguments of its supporters and its critics. Energy sources can be divided into those that are nonrenewable (that is, it took millions of years to create them and they are being used up) and those that are renewable, in the sense that most of them currently gain their energy from the sun, which is expected to continue to shine at its present brightness for at least one billion more years.

Nonrenewable Energy Sources

Oil, natural gas, coal, and uranium are the main sources of nonrenewable energy. Proven reserves of oil in the world in the early 1990s were estimated to be sufficient to allow production at 1989 levels for about 40 years.[56] The production of oil in the United States peaked around 1970 and has been decreasing since then. Proven reserves of natural gas are estimated to be large enough to allow production at 1987 levels for about 55 years.[57] More oil and natural gas will be found in the future, of course, but it is difficult to estimate its quantity or its price.[58]

Coal is a much more abundant resource than oil or natural gas, and the United States has very large deposits of it, as do Russia, China, and Europe. Although a switch back to coal in the United States has not been as fast or as strong as many predicted, many energy analysts expect that this fuel will play a prominent role in the transition period. But the use of coal presents many problems, some of which resulted in the switch to oil after World War II. I was reminded of one of coal's main disadvantages – its polluting effect – when I took my young son to see an old coal-burning passenger train make an excursion run through our town. The black clouds of smoke from the locomotive brought back memories from my childhood when trains like this one passed through our town regularly. Those black clouds dramatized one of the great improvements that came with the arrival of the relatively clean diesel oil trains. Although improvements in technology, such as the use of expensive "scrubbers" on electric generating plants that burn coal, can remove many of the dangerous pollutants caused by burning coal, other pollutants remain. And no affordable technology can prevent the build-up of carbon dioxide in the atmosphere, which comes from burning coal and other fossil fuels.

In addition, over half of the coal in the United States is west of the Mississippi River, in deposits relatively close to the surface. The preferred way of extracting such deposits is by strip mining, a process that wreaks havoc on the land. Efforts to reclaim the land afterward are very expensive and very imperfect. Much of the coal that lies in the eastern part of the country can be obtained only by underground mining, a process that is unhealthy, inherently dangerous, and filled with labor-management conflict.

The making of synthetic oil and gas from coal and oil shale has been considered by some as the way to make the United States less dependent on foreign oil. A multibillion-dollar effort to develop energy this way was approved by Congress in 1980. Vast oil shale deposits lie in Colorado and Wyoming, but the processing of oil shale relies on strip mining and requires vast amounts of water, which is scarce in the West.

113

Efforts to develop this potential source of energy collapsed in the mid-1980s as several large US oil companies withdrew from these ventures because of high construction costs, high interest rates, lowered demand for energy throughout the country, and falling imported oil prices.

Another way to make a synthetic fuel is called coal gasification. Coal can be converted into a synthetic natural gas. The advantage of this procedure is that the process removes nearly all the sulfur that causes acid rain and the synthetic natural gas releases less CO_2 than coal, although natural gas releases even less CO_2 than the synthetic variety. Like the oil shale projects, this technique did not prove to be an economic success since the prices of natural gas and oil fell rather than increased as expected. The US government ended all of its financing for synthetic fuel projects in 1985.

Canada has large deposits of tar sands from which oil can be extracted. But in the mid-1980s Canada too was experiencing serious difficulties developing these deposits as several major oil companies withdrew from the development, leaving the Canadian Government the sole developer in some cases.

It is possible that uranium, the basic fuel for nuclear energy, is widely distributed around the world, but the bulk of positively identified deposits are located in a relatively few countries, one of which is the United States. The mining of uranium can, and has, led to cancer and the waste products from the mining are radioactive. The United States has fairly abundant supplies of uranium, but, like coal, they will eventually run out.

Renewable Energy Sources

The energy from the sun can be obtained in a variety of ways: from wood, falling water, wind, wastes, and, of course, from direct sunlight. We will briefly examine each of these.

First, wood, agricultural and forestry residues, and animal dung are still the principal fuels in developing countries. Rural peoples in sub-Saharan Africa as in the South Asian countries of India, Pakistan, and Bangladesh use these traditional fuels to cook their food and to provide heat and light. In fact, except for their own muscle power and the aid of a few domestic animals, the majority of the 2.5 billion villagers in the developing nations have no other source of energy.[59] Rapidly expanding populations in the Third World are placing high demands on the use of wood; at the same time, modern agricultural requirements and development in general are leading to the clearing of vast acres of forests. Acute shortages of firewood already exist in wide areas of Africa, Asia, and Latin America. The US Government's *Global 2000 Report* projects that the demand for wood for fuel will exceed supplies by about 25 percent before the year 2000.[60]

Second, hydroelectric power, which is generated from falling water, is a clean source of energy, causing little pollution. A large potential for

The destruction of forests for development purposes in the Third World is occurring at the same time as the growing demand for wood as fuel. (*Caterpillar Company*)

developing this type of energy still exists in Africa, Latin America, and Asia, although many of the rivers that could be used are located far from centers of population. Large dams, which are often necessary to store the water for the electric generators, usually seriously disturb the local environment, sometimes require the displacement of large numbers of people, and cause silting behind the dam, which limits its life. While most of the best sites for large dams in the industrialized countries have already been developed, a potential exists for constructing some small dams and for installing electric generators at existing dams that do not have them.

Third, wind is an energy source that was commonly used in the past for power as well as for the cooling of houses. It is still used for these purposes in some Third World countries and has recently gained new respect in the United States, especially in California, where 16,000 wind turbines have been constructed in just three mountain passes, areas that have fairly steady wind. Actually, the midwestern states in the United States – from North Dakota to Texas – have better wind conditions than California and have a great potential for generating more of their power from this source. The dominance of California in producing wind power in the United States probably has more to do with the tax incentives that the state gave in order to promote this form of energy rather than wind conditions.[61] The California wind farms began going up in 1981 after the federal government passed a law that encouraged small energy producers, and after both the federal government and the state of California gave tax credits to the wind producers . (Both federal and state tax credits ended in 1985.)

Wind turbines in Altamont Pass, California. (*US Department of Energy*)

A potential exists for the wider exploitation of this source of energy, and modern technology is producing more efficient wind collectors. The cost of producing electricity from wind has fallen to such a degree – because of the advances in wind turbine technology – that wind power is now competitive with power produced from fossil fuels. Attracted by the success of wind farms in California, a number of European countries such as Britain, Germany, the Netherlands, Italy, and Denmark are planning to increase their wind power. The main problem with wind, of course, is that it is usually not steady, and thus the energy it creates must be stored in some way so it can be used when the wind dies down. There is not yet any easy and inexpensive way to do this. Another problem with wind is that the choice windy places in the world are relatively few and unevenly distributed. They are also often in remote locations, far from population centers, and in areas of great natural beauty, which the windmills spoil. A past problem, the noise the wind turbines make as the blades whirl, has been partly solved by improvements in turbine design.

Fourth, biomass conversion is the name given to the production of liquid and gaseous fuel from crop, animal, and human wastes; from garbage from cities; and from crops especially grown for energy production. Millions of generators that create methane gas from animal and human wastes are producing fuel for villages in India and China. Brazil is using the residues from its large sugar-cane production to produce alcohol for fuel for automobiles and is experimenting with growing cassava, a common root crop, for converting into alcohol. St Louis and some other US cities are burning their garbage mixed with coal and/or natural gas to produce electricity. It is difficult to estimate how widespread this form of energy generation will become in the future. Some see good potential while others mention its negative aspects, such as the emission of harmful gases and of foul odors from burning garbage, and the use of land to grow energy crops instead of food crops in a hungry world.

Fifth, the use of direct sunlight probably has the greatest potential of all the forms of solar energy for becoming a major source of energy in the future. Each year the earth receives from the sun about ten times the energy that is stored in all of its fossil fuel and uranium reserves Direct sunlight can be used to heat space and water, and to produce electricity, indirectly in solar thermal systems, or directly by using photovoltaic or solar cells. Solar thermal systems collect sunlight through mirrors or lenses and use it to heat a fluid to extremely high temperatures. The fluid heats water to produce steam, which is then used to drive turbines to generate electricity. The solar cell was developed for use in the US space exploration program, and extensive research is now being carried on to adapt it for private use.[62]

Solar cells are now being used in calculators and watches. The cells are still relatively expensive, however, and their high cost is probably the most serious hindrance to their wider use. A major reduction in their cost would probably come about if they were mass-produced, but without a large demand for solar cells, which their high cost prevents, mass production facilities will not be built by private enterprise. A way out of the vicious cycle could come through large purchases by the government, which one of solar energy's strong supporters – Barry Commoner – has advocated;[63] or it could come as the costs of other types of energy continue to rise. Solar energy could be used well in moderately or intensely sunny places. Much of the Third World fits this criterion. The Third World is, in fact, often mentioned as a vast potential market for solar energy because many of its rural areas still lack electricity, and solar energy is collected about as efficiently by small, decentralized collectors as it is by larger, centralized units.

The cost of solar energy from solar thermal plants has been dropping rather rapidly. The cost of electricity from these plants is now approximately equal to that produced by the latest nuclear power plants.[64] The World Bank predicts that the cost of solar energy in sunny areas will eventually – over the long term – become competitive with the cost of energy produced by fossil fuels.[65] If one includes the hidden costs of

117

Solar energy provides power for a water pump in Morocco. (*USAID Photo Agency for International Development*)

fossil fuels – i.e. the costs society bears because of the pollution they produce and the costs of military forces to ensure access to them – solar energy might be less expensive than fossil fuels right now.

Sixth, geothermal energy, heat that is produced within the earth's interior and stored often in pools of water or in rock, or as steam under the earth's cool crust, is not a form of solar energy but it is a renewable form of energy. Iceland uses this form of energy to heat many of its homes, and Russia and Hungary heat extensive greenhouses with it. Two US cities, one in Oregon and one in Idaho, use geothermal energy, and a geothermal power plant that produces electricity has been built in northern California. In the early 1990s geothermal energy provided New Zealand and Kenya with nearly 10 percent of their electricity, the Philippines with nearly 20 percent, and El Salvador with nearly 40 percent.

For a few favorable locations in the world, geothermal energy can be utilized, but it is not expected to have a wider potential.

Finally, hydrogen-powered fuel cells have the potential to become a major nonpolluting and efficient source of energy for vehicles. In fuel cells hydrogen and oxygen supplied from the air are combined at low temperatures to produce electricity, which is used to run an electric motor. The vehicles powered by the electric motor would be clean, quiet, highly efficient, and relatively easy to maintain. Hydrogen can be obtained from water by a process that itself uses electricity. If the electricity used to make hydrogen comes from renewable sources such as solar power, wind power, or biomass conversion, hydrogen fuel cells are a renewable source of energy. In the mid-1990s fuel-cell technology was still experimental and hydrogen was much more expensive than gasoline. With the increased enforcing of air pollution standards in a number of US urban areas beginning in the early 1990s, it is possible that a growing demand for vehicles that produce no pollution (called "zero emission vehicles") could lead to improved fuel-cell technology and lower costs.

Conservation

Conservation is not commonly thought of as an energy source, but according to a study of the US energy situation by the Harvard University Business School, it should properly be regarded as a major untapped source of energy. The Harvard study concludes that conservation, rather than coal or nuclear energy, is the major alternative to imported oil.[66] How much energy could the United States save by adopting conservation measures? The Harvard study found that the savings could be impressive:

> If the United States were to make a serious commitment to conservation, it might well consume 30 to 40 percent less energy than it now does, and still enjoy the same or an even higher standard of living. That saving would not hinge on a major technological breakthrough, and it would require only modest adjustments in the way people live.[67]

To many people, the term "conservation" means deprivation, a doing without something; but the Harvard study, and many others, have shown that much energy conservation can take place without causing any real hardship. There are three ways to save energy: by performing some activity in a more energy-efficient manner (for example, design a more efficient motor); by not wasting energy (turn off lights in empty rooms); and by changing behavior (walk to work or to school).

Many US businesses now recognize that making their operations more energy-efficient is a good way to increase profits. The investments they make to improve their business operations and reduce their energy usage are soon repaid by reduced energy bills. One of the most impressive examples is IBM, which set out in 1973 to reduce its energy use by 10 percent and ended up by reducing it by 45 percent.[68] At Dow Chemical, it was discovered after the 1973 oil crisis that the company's standard practice up to then was never to turn off its de-icing equipment during the year, which meant that its sidewalks and service areas were being warmed even on the Fourth of July. US industry has made impressive gains in reducing its energy consumption, but much still can be done.

One major conservation method US industry could adopt is called "cogeneration," which is the combined production of both electricity and heat in the same installation. Electricity is currently produced by private and public utilities, and the heat from the generation of the electricity is passed off into the air or into lakes and rivers as waste. In cogeneration plants, the heat from the production of electricity – often in the form of steam – is used for industrial processes or for heating homes and offices. The production of electricity and steam together uses about one-half the amount of fuel as does their production separately.[69] Cogeneration is fairly common in Europe but not in the United States, where electric utilities often give cheaper rates to their big industrial customers, thus reducing the incentive to adopt the process.

If the United States ever does reach the goal of energy savings that the Harvard report believes is possible, it will be because of a combination of governmental policies encouraging conservation and of action by millions of individuals. The United States is a country where people respond well to incentives to promote conservation practices, but such governmental incentives have so far been rather weak. For several years the federal government allowed individuals who insulated their homes or installed a solar water heater to deduct 15 percent of the cost of the energy improvement from their income taxes. The state of California allows homeowners to deduct 55 percent of the cost of solar devices from their state taxes. (This law no doubt partly explains why California leads the nation in the number of solar devices installed in homes.) The city of Davis, California, has changed its building code so that all new homes in the city must meet certain energy performance standards.

Nevertheless, American homes are not designed to use energy efficiently. If houses with large window surfaces were positioned to face the south, they could gain much heat from the low winter sun, and these windows could be shaded by deciduous trees or an overhang to keep out the high summer sun. The popular all-glass American skyscrapers built during the 1960s are huge energy wasters, since their large areas of glass absorb the hot summer rays. Since their windows cannot be opened, at times the buildings' air conditioners must work at high levels just to cool their interiors to the same temperature as the outside air. Simple measures like planting trees to obtain shade can have a significant cooling effect on a house, a city street, or a parking lot – reducing temperatures by as much as 10 to 20 degrees over unshaded areas. Townhouses, the modern name for the old row houses, are again becoming popular in many cities; they are much more energy-efficient than the common, single family house because of their shared walls. According to a study by the American Institute of Architects, "improved design of new buildings and modification of old ones could save a third of our current total energy use – and save money too."[70]

Conservation: The Case of the Inexpensive Water Heater

A personal blunder illustrates well several conservation principles. A few years ago my hot water heater stopped working. I was greatly relieved when the plumber assured me that he could replace it right away so we could soon have hot water again. I remembered that the government had a policy of rating appliances for their energy efficiency, and asked the plumber if the water heater he would install was energy-efficient. "Oh yes," he replied, "it has a sticker on it." I was also pleased to hear that the price of the new heater was lower than I feared it might be. After the new heater was installed, I read the label on it, which rated its use of energy. I found to my dismay, and embarrassment, that the heater had the lowest rating for energy efficiency that it was possible to give on that model. A water heater lasts about 15 years, so I will have to pay for my error for some time. (Wrapping an insulation blanket around the water heater later helped to reduce the heater's deficiencies.)

In addition to teaching me that conservation requires careful planning, this little episode illustrated well to me a key aspect of energy conservation and explained why many people don't do it. Conservation often requires an initial investment – the more efficient water heaters are more expensive than the least efficient – and the decision to spend more now in order to save in the future is not always easy to make. People naturally look at the purchase price of the appliance – or home – and often follow the rule, "the cheaper the better," as long as the appliance or home is adequate. What that price does not tell you – but what the government's sticker on my water heater did tell me – is how much it will cost to run that appliance, or heat and air-condition that home, over its many years of use.

Saving energy often takes an initial investment, as the box "Conservation: The Case of the Inexpensive Water Heater," illustrates. Knowing this fact helps one understand why the decontrol of prices of oil and natural gas, which will lead to higher prices of those fuels, is probably not enough by itself to cause many people to use less energy. The better educated and more affluent might recognize that an investment in insulation or a more expensive water heater makes good sense and will save them money over the long run, but those with lower incomes do not have the extra money to make the initial investment. Some of the poor spend a higher portion of their income on energy than do those on higher incomes, and thus could benefit greatly from the better-insulated house or the more fuel-efficient car, but they usually end up with a poorly insulated house and a gas-guzzling car. Higher prices for fuel will probably help to reduce energy consumption, but stronger governmental incentives and regulations, such as substantially higher tax credits for installing insulation and substantially higher fuel efficiency standards for automobiles, could produce a significant movement toward conservation.

Nuclear Power: A Case Study ● ● ● ● ● ● ● ● ● ● ● ● ●

In this final section we shall look closely at nuclear power, which is surrounded by political controversy. The supporters and critics of nuclear power are locked in a battle that could decide the fate of this form of energy. We will first look at a brief history of the development of nuclear power so that we can better understand the arguments each side makes.

The Potential and the Peril

Nuclear power was seen by many in its early years as the answer to the world's energy needs. Its promoters claimed it would be a nonpolluting and safe form of energy that could produce electricity "too cheap to meter." After the destructive power of the atom was demonstrated with the bombing of Hiroshima and Nagasaki, people welcomed the thought that atomic research could also be used for peaceful purposes. The first prototype of a commercial nuclear power plant began operation in the United States in 1957. In 1988 nuclear power was producing about 17 percent of the world's electricity. From 1970 to 1980 there was a 24 percent average annual growth in nuclear power, but from 1980 to 1985 the growth had fallen to 16 percent and from 1985 to 1988 it further declined to 7 percent as fewer nuclear power reactors were built.[71] Figure 4.4 shows the number of nuclear power plants around the world in 1991 which were in operation or under construction. Figure 4.5 shows the percentage of the electricity in various countries that was produced by nuclear power in 1991.

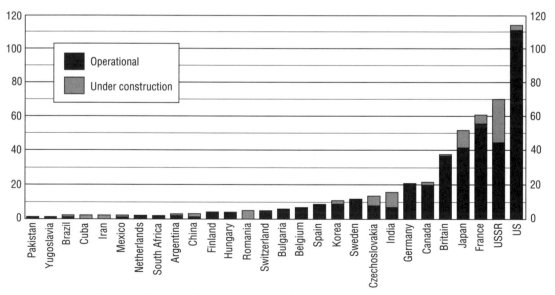

Figure 4.4 Number of nuclear power plants in 1991.

Source: International Atomic Energy Agency.

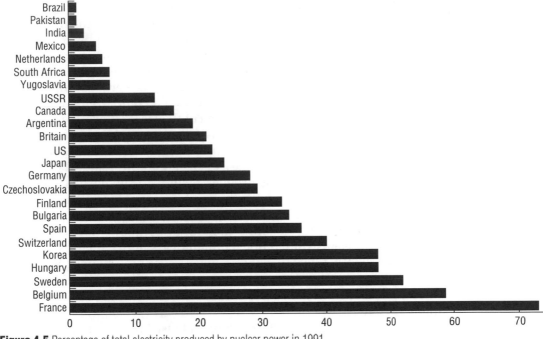

Figure 4.5 Percentage of total electricity produced by nuclear power in 1991.

Source: International Atomic Energy Agency.

The first generation of nuclear power reactors is known as fission, light-water reactors. They utilize the same process that was used to explode the early atomic bombs – the splitting of the core (the nucleus) of the atoms of heavy elements, which releases tremendous energy. Uranium 235 (U-235) is the fuel used in these reactors. The chain reaction that comes with the splitting of the uranium nucleus is controlled in the power reactors to produce sustained heat, which is then used, as it is in coal- and oil-fed power plants, to produce steam. The steam drives the turbines that generate electricity.

The uranium used in the common light-water reactors must be enriched so that it contains a percentage of U-235 higher than that found in nature. This is done in very large, very expensive, enrichment plants that utilize huge amounts of electricity themselves. Because of the difficulty of obtaining the required U-235, it was originally planned to reprocess the spent fuel rods from the power reactors to extract unused uranium, thus making uranium supplies last longer. Controversy has surrounded these reprocessing plants, partly because plutonium, one of the deadliest known substances and the fuel for the Nagasaki bomb (the bomb dropped on Hiroshima utilized uranium), is produced during the reprocessing. Three commercial reprocessing plants were built in the United States – one in Illinois, one in New York, and one in South Carolina – but none are operating at present. The shutdowns occurred because of technical difficulties, safety concerns, and the fear

123

that such plants made plutonium too accessible.

Another way to handle the relative scarcity of fuel for the light-water reactors would be by building a second generation of power reactors. These reactors, known as the fast-breeder reactors, use plutonium as their fuel and will actually produce more fuel than they consume. The attractiveness of this feature was countered by the great complexity of the plants and the increased danger that would come from an accident, since plutonium was being used instead of the less radioactive uranium. Plutonium is extremely harmful if inhaled or digested and has a half-life (the amount of time for one-half of the substance to disintegrate or be transformed into something else) of over 24,000 years. The Soviet Union began operating a fast-breeder reactor in the early 1970s, and India completed construction of an experimental breeder reactor in the mid-1980s. After spending nearly $2 billion on an effort to build a demonstration breeder reactor, the US government abandoned the effort in 1983 because of technical difficulties, safety concerns, and spiraling costs. In 1984 five European countries (Britain, France, West Germany, Belgium, and Italy) agreed to pool their efforts to build breeder reactors, although Britain later withdrew from the agreement for economic reasons.

Japan has built two small experimental fast-breeder reactors. In the early 1990s it was building a large demonstration breeder reactor and a reprocessing plant for the recycling of spent uranium fuel and for the production of plutonium. Its plans to eventually build a number of breeder reactors and to ship many tons of plutonium from reprocessing facililties in France and Britain for the reactors to use were being reassessed in the early 1990s because of economic reasons and opposition by a number of foreign countries. Although few doubted Japanese current assurances that the plutonium would be used for civilian purposes, there was a fear that in the future the plutonium could be used to build nuclear weapons. There was also concern about the safety of shipping the plutonium to Japan and and its storage in Japan. With the price of uranium relatively low and supplies plentiful, Japan began rethinking its prior decision, made during the energy crises of the 1970s, to rely on plutonium-fed breeder reactors to gain independence in energy production.[72]

Fusion nuclear power, which is still in the experimental stage, might be called a third generation of nuclear energy. Fusion energy is created by the same process that creates the energy in the sun and is the process used in the hydrogen or H-bomb, which is vastly more powerful than the fission atomic bombs. Instead of splitting atoms, as happens in fission, in fusion atoms are fused together. The process is highly complicated and demands temperatures (millions of degrees) and pressures with which scientists have little experience. The attractiveness of the fusion process is that it is an inherently safer process than fission and it generates much less radioactive waste. Much of its fuel (deuterium) comes from seawater and is nonradioactive, while the other main component of its fuel (tritium), which is radioactive, can be obtained from a substance (lithium) that is fairly abundant.

In 1991, after nearly a half century of research and many billions of dollars, a breakthrough in fusion research occurred as a European team for the first time produced a significant amount of energy from controlled nuclear fusion. Four industrial powers (Europe, Japan, the Commonwealth of Independent States, and the United States) have agreed to cooperate in building a large experimental fusion reactor. The reactor is scheduled to be completed early in the twenty-first century. If experiments in fusion continue to be successful, a fusion reactor power station could begin operation for demonstration purposes around 2020.[73]

Nuclear power is in serious difficulty around the world, with only France and Japan aggressively promoting it. China built its first nuclear power station in the early 1990s and plans to build more to provide power for its coastal provinces where industrialization is spreading rapidly and major energy shortages exist. In many Western European countries there has been little new construction and few orders for new plants. The former Soviet Union had ambitious plans to expand nuclear power, but after the accident at Chernobyl, the critics of nuclear power have forced the new governments of the former Soviet republics to delay new construction.

There is now a major concern about the continued safety of many of the nuclear plants in the Eastern European countries, which were part of the Soviet empire before it collapsed. There is concern also for many plants in Russia and in other former Soviet republics.[74] Many of these reactors have poor maintenance and design problems – such as no containment structures to prevent radioactive material from escaping the plants in the event of an accident and poor equipment to cool the core in case of an emergency – but because of serious financial difficulties that came after the collapse of the Soviet Union, funds are not available to repair or improve the reactors.

In the United States there have been no new orders for nuclear power plants since 1977, and about 100 orders for plants were canceled during the 1970s. This slowdown has occurred because of the reduced demand for electricity (caused by the rapidly increasing cost of power, by conservation measures, and an economic recession), the skyrocketing cost of building the plants (plants that were originally estimated to cost from \$200–\$300 million now cost from \$1–\$2 billion), and increasing concern about the safety of the plants.

In 1979 the partial meltdown of the core of the nuclear reactor located at the power plant at Three Mile Island, Pennsylvania, led to a release of some radioactive steam and gas from the plant and the consequent official recommendation that nearby pregnant women as well as young children be evacuated. Although no one was killed, and the release of the radioactive substances was later judged by a presidential investigation committee to have caused no danger to public health,[75] the cleanup from the accident cost about \$1 billion and public fears about

125

nuclear power increased. It has been estimated that new safety requirements for nuclear power plants that were issued by the federal government after the Three Mile Island accident, as well as delays in
construction of new plants and additional temporary shutdowns of
existing plants, caused by the concerns raised by the Three Mile Island
accident, added $130 billion to the cost of nuclear electricity in the
United States between 1979 and 1992.[76] Two lessons learned from the
accident were that it was easier to destroy a reactor core than many
experts had thought possible and that it was harder to rupture the reactor vessel (the five-inch-thick steel pot that holds the core) than many
had thought possible.[77]

In 1986, when the nuclear power plant at Chernobyl in the Soviet
Union exploded, about 50 tons of radioactive particles – ten times the
fallout at Hiroshima – fell across parts of the Soviet Union. The accident
also spread radiation around the world, with significant amounts falling
on some European countries. More than 100,000 people were evacuated from an area of about 300 square miles (78,000 hectares) around
the plant. About 30 people died from the catastrophe during the following few months. It is believed that both flaws in the design of the reactor
and mistakes by the operators of the plant were responsible for the disaster.

While the Soviet Union was much more candid about the accident
than it had been about previous nuclear diasters (such as the explosion,
which was kept secret, in 1957 of a tank that contained high-level
nuclear waste and contaminated thousands of square miles in the Ural
Mountains area), it is now known that secret government decrees were
issued one and two years after the accident, designed to cover up the full
extent of the damage.[78] Four years after the accident, the Soviet Union
acknowledged that 4 million people were still living on ground contaminated by the explosion and the goverment voted to spend $26 billion on
further Chernobyl-related expenses, including the resettlement of
200,000 people living in the most contaminated areas.[79] Western
experts differ about the eventual toll from Chernobyl – two US experts
give estimates ranging from 1,000 to 7,500 new cancers over the next 50
years to an estimate of 50,000 to 250,000 related deaths in the former
Soviet Union – but some experts believe that the toll outside the former
Soviet republics will equal that from within.[80]

The Choice

There are two basic political alternatives that exist regarding nuclear
power: the government can withdraw its support for the nuclear power
industry, or it can continue to promote nuclear power and encourage its
development. We will examine the main arguments being presented in
this debate.

Withdraw Support for Nuclear Power

As the accident at Chernobyl shows, nuclear power plants run the risk of having catastrophic accidents. Society should not have to accept such a risk. The Three Mile Island accident proved that the interaction of human error and failure of equipment can lead to events that no one had ever guessed could happen. One hundred alarms went off in the first few minutes at Three Mile Island, making it impossible for the operators to control the situation. And at Chernobyl a series of errors by the operators of a reactor that was inherently difficult to operate generated forces and effects that up to then had never been experienced. Nuclear technology assumes better human performance and understanding than history shows can be achieved.

Nuclear power also increases the danger of the proliferation, or the spread, of nuclear weapons to new nations. The knowledge and installations that a nation acquires when it develops nuclear power can be utilized to develop nuclear bombs. While it is true that most nuclear power plants do not use fuel that can readily be used to build bombs, the reprocessing of the spent fuel from the plants can produce fuel for weapons. Reprocessing technology is spreading around the world. For example, in 1975 West Germany sold Brazil reprocessing technology even though Brazil had never signed the Nuclear Nonproliferation Treaty, which commits nations not to develop nuclear weapons. As more nations gain expertise in nuclear energy and acquire nuclear weapons, chances that the weapons will be used will thus greatly increase.

The danger of terrorists stealing plutonium to make a bomb or to use as a poison to be spread in the atmosphere over a city or in its drinking water is real. The knowledge of how to make nuclear weapons is widespread, and only about 20 pounds of fissionable material is required to make a crude bomb. If, or probably more realistically when, terrorists do acquire a nuclear capability, threatened nations will probably respond by giving their police and governments increased power. The United States could become more authoritarian because of such a threat, with a consequent restriction on personal freedom and privacy.

Another disturbing consequence of nuclear energy is that large amounts of radioactive wastes from nuclear power plants are accumulating in the United States because no permanent way to store this material has yet been created. Some nuclear waste must be stored for 100,000 to 200,000 years in such a way that it does not come in contact with humans or with any part of the environment. We show an overwhelming arrogance and lack of concern for future generations when we say we can do this. This is made apparent when we remember that the United States is only 200 years old and human civilization about 5,000 years old. Several hundred thousand gallons of nuclear wastes have already leaked out of steel tanks at sites in Hanford, Washington, and near Aiken, South Carolina, where military wastes are temporarily stored.

The high cost of nuclear power makes it economically unjustifiable.

If one includes the cost of its development and the costs of the attempts to find a safe way to store wastes – both of which have been financed by public funds – nuclear power would not be competitive with other forms of power. And the cost of decommissioning and possibly hauling away for storage worn out and highly radioactive nuclear power plants, which have a life expectancy of 30 to 40 years, will further add to the costs of nuclear power.

Nuclear power plants create thermal pollution, raising the temperature of the atmosphere and of the water in lakes, bays, and rivers that are often used to cool the reactors. The warmer water is deadly to many kinds of fish and other forms of life in the lakes and rivers. The present light-water reactors convert only about 30 percent of their fuel into electricity; the rest is turned into waste heat.

Nuclear power cannot replace imported oil. Only about 10 percent of the world's oil is used to generate electricity; most of it is used to run vehicles, provide heat for homes and industry, and make chemicals. Nuclear power, for the foreseeable future, can be used only to make electricity, and, besides being of relatively limited use, electricity is a very expensive form of energy. The huge public investments that continue to go into nuclear energy prevent public funds from being used to develop alternative sources of energy that would be more useful, safer, and cleaner.

Continue to Support Nuclear Power

We accept chemical plants despite accidents associated with them, such as that at Bhopal, India, in 1984 which killed more than 2,000 people, and we accept dams despite accidents associated with them, such as at the Vaiont Dam in Italy in 1963, which also killed about 2,000 people. So why should the accidents at Chernobyl and Three Mile Island make us reject nuclear power? Nuclear power plants in the United States have become much safer because of new procedures and safety devices adopted since the Three Mile Island accident. And a number of highly respected scientists now believe that it is possible to build a nuclear power plant that is inherently safe, one that would be so designed that if anything unusual happened it would automatically cease functioning without any action needed by human beings or by machines.

A person gets more exposure to radiation in a year by taking a single round-trip coast-to-coast jet flight, by watching color television, or by working in a building made of granite than she or he would by living next to a nuclear power plant. Humans have evolved over millions of years living on a mildly radioactive planet and have prospered. Few things in life are risk-free, and the risks associated with nuclear power are relatively benign compared with the risks people take every day in their lives.

The accusation that nuclear power will contribute to the proliferation of nuclear weapons is exaggerated. A nation that wants to build a nuclear weapon can get sufficient plutonium from its nuclear research facilities. This is exactly how India got plutonium for the nuclear device it exploded in 1974. All of the major powers that have acquired nuclear weapons built these weapons before they acquired nuclear power.

It would be very difficult for a terrorist to steal plutonium in the United States. The US military has been shipping plutonium by convoy for many years and very effective means have been devised to protect the shipments from hijacking. Good security measures are also in effect in plants that produce plutonium. Although the knowledge of how to construct nuclear bombs is no longer secret, the actual construction of such a device is very difficult. If the construction of nuclear weapons were easy, more nations would have them than the handful that do at present.

A way to store nuclear wastes permanently has been devised and is actually being used in Europe. The wastes can be solidified, usually in glass, and then stored in geologically stable underground facilities. France has adopted this method, while other European countries have developed other permanent methods. Most of the nuclear wastes in the United States are being created in its military program, and these wastes will continue to build up even if every nuclear power plant should be closed.

A great danger to the world is caused by the shortage of oil. Without secure sources of energy, such as nuclear power, it is likely that more wars will occur as nations fight to keep their sources of oil secure. The development of nuclear power can help to reduce this dangerous dependency on foreign sources of energy.

Nuclear power is much less polluting than the main alternative energy source – coal – which will be greatly expanded if nuclear power does not continue to be produced. Nuclear power produces no waste gases causing acid rain, as coal burning does, it does not contribute to the build-up of carbon dioxide in the atmosphere, and it does not produce smog or any of the other harmful effects commonly associated with coal. Nuclear power is generally much easier on the landscape than is coal; the average nuclear power reactor uses only about 30 tons of fuel per year while the average coal-burning electric plant uses about 3,000 tons of fuel per day. By slowing down the approval of new nuclear plants, the critics of nuclear power are causing nations to burn more coal. This causes many more people to die and more environmental damage from pollution than would have been the case if new nuclear plants had made the increased coal-burning unnecessary.

No energy option should be rejected during this period of transition. Nuclear power is one of the few alternatives we have to produce large amounts of energy during the rest of this century while the search for a sustainable fuel to take the place of oil continues.

Conclusions

Although the price of oil dropped unexpectedly in the mid-1980s, the long-term prospects for oil remain unchanged. As a nonrenewable resource on a planet with a growing population and expanding industrialization, it is a source of energy that will become more scarce and more expensive, especially as efforts are made to control the pollution it produces. Development in the Third World that rests on a petroleum base is as open to disruption as it was in the 1970s when two oil shocks sent prices shooting upward. Those shocks helped begin the debt crisis that many Third World nations face today, as they borrowed to help pay for their higher bills for imported oil. This was a clear warning to the developing nations that their plans to industrialize along the lines followed in the West were based on a faulty assumption – namely, that oil would continue to be available at a low price.

A long-term strategy for the improving of living standards in many poorer nations would seem to call for the development of sources of energy more secure and less costly than oil. Dependency on oil can be reduced by the adoption of new energy-efficient manufacturing processes. Renewable energy sources such as wood, hydroelectric power, biomass conversion, and solar power offer opportunities for expansion in many Third World nations. The challenge facing the developing nations is to create needed economic growth without increasing their dependency on oil.

The United States responded fairly well to the increased costs of oil, at least in the short run. The role of prices worked just as the market approach said it would in a capitalist economy. Many industries in the United States gave new attention to ways to make their use of oil more efficient, and

succeeded to such a degree that their reduced consumption of oil was one of the factors behind the falling oil prices. New policies by the government also were effective in reducing the demand for oil, such as the required higher fuel-efficiency standards for new automobiles and tax benefits for installing energy-efficient insulation in homes. And especially impressive was the response by millions of individuals in the country who gave up their preference for large, gas-guzzling automobiles and bought small fuel-efficient vehicles. Urgent calls by high government officials for Americans to conserve energy had much less effect than did higher prices for gasoline and fuel oil.

But prices work both ways, and a danger exists that lower oil prices in the 1990s will cause many Americans to return to their more energy-profligate ways. One mitigating factor is that many energy conservation actions already taken cannot be easily reversed. A home insulated will remain insulated no matter what the attitude of the owner toward energy is now. Another factor that continues to promote conservation in the United States is that, although the price of oil is lower than expected, the price of electricity continues to rise as the huge cost of new nuclear power plants is passed on to consumers.

Disturbing signs exist when one looks at the long-term energy plans of the United States. No energy plan exists to reduce the dependency of the country on foreign oil. The nuclear energy industry was in the doldrums even before the disaster at Chernobyl. Governmental support for research in solar and other renewable energy sources all but disappeared in the 1980s under the Reagan and Bush administrations, although a new administration in the early 1990s was showing a renewed interest in renewable energy. The recommendation that a tax be placed on imported oil to keep up the incentive to conserve oil was politically unpopular.

In short, the efforts by the leading industrial nation are not impressive in moving forward in this period of energy transition toward an economy based on renewable and non-polluting sources of energy. The needed political leadership in this crucial matter has not yet appeared, which may reflect a lack of awareness by the public that new energy initiatives are needed for the long-term health of the country.

Notes

1 An interesting discussion of the laws of thermodynamics and what they tell us about energy is contained in Jeremy Rifkin, *Entropy: A New World View* (New York: Viking Press, 1980), pp. 33–43.

2 *New York Times*, national edn. (January 16, 1992), p. A7; and Youssef Ibrahim, "Gulf War's Cost to Arabs Estimated at $620 Billion," *New York Times*, national edn (September 8, 1992), p.A4.

3 Joseph Romm, "Needed — A No-Regrets Energy Policy," *The Bulletin of the Atomic Scientists,* 47 (July/August 1991), p. 31.

4 "Armed Force and Imported Resources," *The Defense Monitor*, 11, no. 2 (1992), p. 1.

5 Lester R. Brown, *The Twenty-ninth Day* (New York: W. W. Norton, 1978), pp. 205-6.

6 Daniel Yergin and Martin Hillenbrand (eds), *Global Insecurity: A Strategy for Energy and Economic Renewal* (Boston: Houghton Mifflin, 1982), p. 7

7 Council on Environmental Quality and the Department of State, *Global Future: Time to Act* (Washington: Government Printing Office, 1981), p. 46.

8 Matthew Wald, "Gas is at Historic Low, Reckoning for Inflation," *New York Times*, national edn (July 26, 1991), p. C1

9 Harold M. Hubbard, "The Real Cost of Energy," *Scientific American* 264 (April 1991), p. 36.

10 Lester C. Thurow, *The Zero-sum Society: Distribution and Possibilities for Economic Change* (New York: Basic Books, 1980), p. 26.

11 Yergin and Hillenbrand, *Global Insecurity*, p. 11.

12 This analysis is based mainly on that made by Daniel Yergin in Yergin and Hillenbrand, *Global Insecurity* , pp.11-12.

13 David Sanger, "Japan Joins in Embargo Against Iraq," *New York Times*, national edn (August 6, 1990), p. C7.

14 David Sanger, "Japan's Oil Safety Net: Will it Hold?" *New York Times*, national edn (September 9, 1990), p. C1.

15 John H. Gibbons, Peter D. Blair, and Holly L. Gwin, "Strategies for Energy Use," *Scientific American*, 261 (September 1989), p. 142.

16 James Tyson, "China Turns On to Nuclear Power," *Christian Science Monitor* (March 25, 1992), p. 12.

17 *New York Times*, late city edn (November 10, 1981), p. C1.

18 Vaclav Smil, *Energy in China's Modernization: Advances and Limitations* (Armonk, New York: M.E. Sharpe, 1988), p. 49.

19 Sheryl WuDunn, "Hopes Fade as Output of Oil Lags in China," *New York Times*, national edn (September 3, 1990), p. 28.

20 "World Bank Warns on Third World Debt," *New York Times*, national edn (December 16, 1991), p. C2.

21 "Brazil Debt Payment Set," *New York Times*, national edn (March 27, 1992), p. C16. One way Brazil hopes to reduce the amount of oil it imports is by using alcohol to fuel its cars. It is using its huge sugarcane wastes to produce alcohol and mix this with its gasoline. The country plans eventually to use only alcohol in its automobiles.

22 Wolfgang Sassin, "Energy," *Scientific American*, 243 (September 1980), p. 121; *New York Times*, late city edn (October 21, 1981), p. 1.

23 United Nations, *Statistical Yearbook*, 1988/89 (New York: United Nations, 1992), table 118.

24 Most of the US consumption of oil now goes for transportation.

25 The Thunderbird continued to grow in 1975 even though there had been a substantial increase in the price of gasoline. This is explained mainly by the fact that the planning and design of new automobiles takes place several years before the car actually appears on the market for sale.

26 Dennis Pirages, *The New Context for International Relations: Global Ecopolitics* (North Scituate, Mass.: Duxbury Press, 1978), p. 112.

27 Matthew Wald, "Where All That Gas Goes: Driver's Thirst for Power," *New York Times*, national edn (November 21, 1990), p. A1.

28 James R. Chiles, "Tomorrow's Energy Today," *Audubon* (January 1990), p. 61.

29 Matthew Wald, "Gulf Victory: An Energy Defeat?" *New York Times*, national edn (June 18, 1991), p. C3; and Matthew Wald, "U.S. Oil Output Drops; Consumption Also Falls," *New York Times*, national edn (January 16, 1992), p. C6.

30 David Sanger, "Japan Joins in Embargo Against Iraq," *New York Times*, national edn (August 6, 1990), p. C7 In the early 1990s Japan's energy intensity (the amount of energy required to produce a unit of gross domestic product) was the lowest of all market industrialized countries. These countries, in the previous two decades, had reduced their energy intensity by 25 percent, with the sharpest decline coming after 1979. World Resources Institute, *World Resources 1992–93*, p. 145.

31 Matthew Wald, "Gulf Victory: An Energy Defeat?" *New York Times*, national edn (June 18, 1991), p. C3.

32 For example, see Marc H. Ross and Daniel Steinmeyer, "Energy for Industry," *Scientific American,* 263 (September 1990), pp. 88-98; Gibbons et al., "Strategies for Energy Use," pp. 136-43; the National Research Council, *Energy in Transition 1985-2010* (San Francisco:

W. H. Freeman, 1980); Robert Stobaugh and Daniel Yergin (eds), *Energy Future: Report of the Energy Project at the Harvard Business School* (New York: Ballantine Books, 1980); Roger W. Sant et al., *Eight Great Energy Myths: The Least-Cost Energy Strategy, 1978-2000* (Arlington, Va.: The Energy Productivity Center of the Mellon Institute, 1981).

33 Council on Environmental Quality, *Global Future*, p. 61.

34 Joel Darmstadter, "Economic Growth and Energy Conservation: Historical and International Lessons," reprint no. 154, (Washington: Resources for the Future, 1978), p. 18.

35 In 1974 the average fuel efficiency of all American cars was 14 miles per gallon. The law required that this be increased to 27.5 miles per gallon by 1985.

36 Nicholas von Hoffman, "'Reinventing' America to Save Energy," *New York Times Magazine* (September 23, 1979), p. 62.

37 Warren M. Washington, "Where's the Heat?" *Natural History* (March 1990), p. 67.

38 Philip D. Jones and Tom M. L. Wigley, "Global Warming Trends," *Scientific American*, 263 (August 1990), p. 91.

39 Richard A. Houghton and George M. Woodwell, "Global Climatic Change," *Scientific American*, 260 (April 1989), p. 40; Fakhri A. Bazzaz and Eric D. Fajer, "Plant Life in a CO_2-Rich World," *Scientific American*, 266 (January 1992), p. 68.

40 Robert M. White, "The Great Climate Debate," *Scientific American*, 263 (July 1990), p. 40.

41 William Stevens, "Two Studies Rank 1990 as the Warmest Year," *New York Times*, national edn (January 10, 1991), p. A1.

42 J. T. Houghton, G. J. Jenkins, and J. J. Ephraums (eds) *Climate Change: The IPCC Scientific Assessment* (Cambridge: Cambridge University Press, 1990), p. xxix.

43 Ibid.

44 Ibid., p. xi. It is expected that by 2025 the global mean temperature will increase about 1° C over the 1990 value. The amount of CO_2 in the atmosphere will not stop increasing after it has doubled, of course. Depending on how much carbon continues to be released on earth, the CO_2 level in the atmosphere could keep rising after the doubling.

45 J. Hansen et al., "Climate Impact of Increasing Atmospheric Carbon Dioxide," *Science*, 213 (August 28, 1981), p. 966.

46 J. T. Houghton et al., *Climate Change: The IPCC Scientific Assessment*, p.xxiv.

47 Ibid., p. xi.

48 See for example, S. Fred Singer, "Warming Theories Need Warning Label," *The Bulletin of the Atomic Scientists*, 48 (June 1992), pp. 34–9, and Philip H. Abelson, "Uncertainties About Global Warming," *Science*, 247 (March 30, 1990), p. 1529.

49 For a description of other possible "feedbacks," see Jon R. Luoma,

"Gazing into Our Greenhouse Future," *Audubon* (March 1991), pp. 57–8.

50 J. T. Houghton et al., *Climate Change: The IPCC Scientific Assessment*, pp. xi and xxvii.

51 Council on Environmental Quality, *Global Energy Futures and the Carbon Dioxide Problem* (Washington: Council on Environmental Quality, 1981), p. 52.

52 Humus, the organic material in topsoil, also stores large amounts of carbon.

53 Wallace S. Broecker, "Global Warming on Trial," *Natural History* (April 1992), p. 14.

54 Council on Environmental Quality, *Global Energy Futures*, pp. vii-viii.

55 World Bank, *World Development Report 1992 – Development and the Environment* (New York: Oxford University Press, 1992), p. 124.

56 World Resources Institute, *World Resources 1992–93*, (Oxford: Oxford University Press, 1992), p. 149.

57 United Nations Environmental Programme, *Environmental Data Report*, (Oxford: Basil Blackwell, 1991), p. 292.

58 The World Bank believes that fossil fuel reserves are sufficient to meet world demands through the twenty-first century, and even perhaps longer. See World Bank, *World Development Report 1992*, pp. 115-16.

59 World Resources Institute, *World Resources 1992–93*, p. 144.

60 Council on Environmental Quality and the Department of State, *The Global 2000 Report to the President: Entering the Twenty-First Century*, vol. 1 (New York: Penguin Books, 1982), p. 2.

61 Matthew Wald, "Putting Windmills Where It's Windy," *New York Times*, national edn (November 14, 1991), p. C1 and Matthew Wald, "A New Era For Windmill Power," *New York Times*, national edn (September 8, 1992, p. C1).

62 Some analysts argue that oil companies, which own most of the patents on solar cells, may be delaying the development of solar energy so that they can maximize their profits from fossil fuels. See Ray Reece, *The Sun Betrayed: A Report on the Corporate Seizure of U.S. Solar Energy Development* (Boston: South End Press, 1979), and Richard Barnet, *The Lean Years: Politics in the Age of Scarcity* (New York: Simon and Schuster, 1980).

63 Barry Commoner, *The Politics of Energy* (New York: Alfred A. Knopf, 1979).

64 James Chiles, "Tomorrow's Energy Today," p. 63.

65 World Bank, *World Development Report 1992*, p. 123.

66 Stobaugh and Yergin, *Energy Future*, p. 10.

67 Ibid., p. 167.

68 Frances Gendlin, "A Talk with Daniel Yergin," *Sierra*, (July/August 1981), p. 62.

69 Stobaugh and Yergin, *Energy Future*, p. 198.

70 Quoted in Barnet, *The Lean Years*, p. 100.

71 United Nations Enviroment Programme, *Environmental Data Report*, pp. 293–4.

72 David Sanger, "Japan Is Cautioned on Plan to Store Tons of Pluto-nium," *New York Times*, national edn (April 13, 1992, p. A2; and David Sanger, "Japan Thinks Again About Its Plan to Build a Pluto-nium Stockpile," *New York Times*, national edn (August 3, 1992), p. A3).

73 Robert W. Conn et al., "The International Thermonuclear Experi-mental Reactor," *Scientific American*, 266 (April 1992), p. 103.

74 Marlise Simons, "East Europe's Nuclear Plants Stir West's Safety Concerns," *New York Times*, national edn (June 7, 1990), p. A1; and Matthew Wald, "U.S. Sees Rising A-Plant Dangers in East Europe and Soviet Union," *New York Times*, national edn (October 8, 1991), p. A4).

75 Dorothy Nelkin, "Some Social and Political Dimensions of Nuclear Power: Examples from Three Mile Island," *The American Political Science Review*, 75 (March 1981), p. 135.

76 *New York Times*, late city edn (May 30, 1986), p. A9.

77 Matthew Wald, "After the Meltdown, Lessons From a Cleanup," *New York Times*, national edn (April 24, 1990), B6.

78 Felicity Barringer, "Chernobyl: Five Years Later the Danger Per-sists," *New York Times Magazine* (April 14, 1991), p. 36.

79 "Four Years Later, Chernobyl Still Claims Victims," *New York Times*, national edn (April 26, 1990), p. A3; and Felicity Barringer, "Four Years Later, Kremlin Speaks Candidly of Chernobyl's Hor-rors," *New York Times*, national edn (April 28, 1990), p. 1.

80 Felicity Barringer, "Chernobyl," p. 39.

Further Readings

Davidson, Art, *In the Wake of the Exxon Valdez: The Devastating Impact of the Alaska Oil Spill* (San Francisco: Sierra Club Books, 1990). This reads like a mystery book as the author reveals the events that led up to this major environmental disaster.

Flavin, Christopher, and Nicholas Lensen, *Beyond the Petroleum Age: Designing a Solar Economy*. Worldwatch Paper No. 100 (Washington: Worldwatch Institute, December 1990). Working from the assumption that an energy transition must occur soon, the authors explore the alter-natives and discuss the resultant policy changes.

Flavin, Christopher, and Alan T. Durning, *The Age of Energy Efficiency*. Worldwatch Paper No. 82 (Washington: Worldwatch Institute, March 1988). An overview of what has been learned about energy efficiency, what is presently within our ability, and what may be possible.

Kapstein, Ethan B., *The Insecure Alliance: Energy Crisis and Western Politics Since 1944* (New York: Oxford University Press, 1990). The author traces the history of the Western alliance as it dealt with the supply of oil and shows how the handling of crises over oil changed over time.

Lensen, Nicholas K., *Nuclear Waste: The Problem That Won't Go Away*. Worldwatch Paper No. 106. (Washington: Worldwatch Institute, December 1991). Lensen focuses on the permanent nature of nuclear waste, the health hazards, and the lack of sufficient knowledge to deal with the problem.

Long, Robert Emmet, *Energy and Conservation* (New York: H. W. Wilson, 1989). An introduction to the issues of fossil fuel conservation and the development of alternative sources.

Lyman, Francesca, et al., *The Greenhouse Trap* (Boston: Beacon Press, 1990). Beginning with an overview of the history of development and atmospheric damage, the authors carefully clarify what global warming means to us, our environment, our food and our health. In addition, they explore the possibilities for climate stability.

Medvedev, Grigori, *The Truth About Chernobyl* (New York: Basic Books, 1991). The author, the head of the official investigation into the causes of this disaster, describes errors in the design, and negligence in the overall administration and day-to-day operations of the plant, which made this accident nearly inevitable.

Morone, Joseph G., and Edward J. Woodhouse, *The Demise of Nuclear Energy? Lessons for Democratic Control Of Technology* (New Haven: Yale University Press, 1989). The authors explore the history of nuclear energy in the United States, including glaring mistakes at Three Mile Island, and suggest that the technology itself is not to blame, but rather that political and economic factors have prescribed a path for development that might not be the best course.

Schneider, Stephen H, *Global Warming: Are We Entering the Greenhouse Century?* (San Francisco: Sierra Club Books, 1989). A leading climatologist offers a detailed look at the effects of global warming, discussing exactly what a slight rise in temperature, for a year or a century, would mean to agriculture, public health, and the environment.

Scientific American, vol. 263 (September 1990). This entire issue is devoted to energy. It includes the following articles : "Energy for Planet Earth," "Efficient Use of Electricity," "Energy for Buildings and Homes," "Energy for Industry," "Energy for Motor Vehicles," "Energy for the Developing World," "Energy for the Soviet Union, Eastern Europe and

China," "Energy from Fossil Fuels," "Energy from Nuclear Power," "Energy from the Sun," and "Energy in Transition."

Yergin, Daniel, *The Prize: The Epic Quest for Money, Oil, and Power* (New York: Simon and Schuster, 1990). Yergin explores three themes: oil has played a central role in the development of modern capitalism and modern business; oil is closely tied to the political strategies of nations and to their efforts to achieve power; and modern society is so dependent on oil that it can be called a Hydrocarbon Society, peopled by Hydrocarbon Man.

5

The Environment

> *We travel together, passengers on a little spaceship,*
> *dependent on its vulnerable resources of air and soil;*
> *all committed for our safety to its security and peace; preserved*
> *from annihilation only by the care, the work, and I will say,*
> *the love we give our fragile craft.*
>
> Adlai E. Stevenson (1900–65)

The Awakening ● ● ● ● ● ● ● ● ● ● ● ● ● ● ● ● ● ● ●

The relationship between the environment and development has not been a happy one. Development has often harmed the environment, and the environmental harm has in turn adversely affected development. Industrialization brought with it many forms of pollution, pollution that is undermining the basic biological systems upon which life rests on this planet. It took millions of years for these systems to evolve.

The first world conference on the environment was held in Stockholm, Sweden, in 1972 under the auspices of the United Nations. At that conference the developed nations, led by the United States, pushed for greater efforts to protect the environment while many less developed nations feared that an effort to create strict antipollution laws in their countries would hurt their chances for economic growth. The developing nations maintained that poverty was the main cause of the deterioration of the environment in their countries. What they needed, they said, was more industry instead of less.

138

Ten years later, the nations of the world again met together to discuss the state of the global environment, this time in Nairobi, Kenya. The positions of the rich and poor nations had changed dramatically. The developing nations generally showed enthusiasm for further efforts to protect the environment, since in the ten years between the conferences they had seen that environmental deterioration, such as desertification, soil erosion, deforestation, and the silting of rivers and reservoirs, was harming their efforts to develop and to reduce poverty. On the other hand, many of the rich nations at Nairobi, led by the United States, called for a slowing down of environmental initiatives until they had recovered from their economic recessions.

Even though the positions of the developed and developing nations had become somewhat reversed during the ten years between the two environmental conferences, there is no doubt that an awareness of the threat to the environment caused by human activities had by 1982 become worldwide. Only 11 nations had had any kind of governmental environmental agency at the time of the first conference, whereas over 100 nations, 70 of them in the Third World, had such agencies at the time of the second . These agencies did much to educate their own governments and people about environmental dangers.

In 1992 the third environmental conference sponsored by the United Nations was held in Rio de Janeiro, Brazil. Popularly called the Earth Summit, and formally the Conference on the Environment and Development, it was attended by the largest number of leaders of nations in history. They were joined by about 10,000 private environmentalists from around the world and 8,000 journalists. Although frequent clashes took place between the representatives of Northern rich countries and relatively poor Southern countries in the preparatory meetings, which took place during the two years preceding the conference, two major treaties were signed by about 150 nations at the conference. One concerned the possible warming of the earth's climate, which was discussed in the chapter on energy. The treaty called on nations to curb the release of so-called greenhouse gases that may be causing a change in the world's climate. Because of the insistence of the United States, no specific targets or timetables were placed in the treaty, but the treaty did call for nations to eventually reduce the emissions of their greenhouse gases to 1990 levels.

The second treaty – the biodiversity treaty, providing for the protection of plant and animal species – was signed by most nations. The United States did not sign it and stood fairly alone in its opposition to it. The opposition by the Bush administration in the United States to these environmental initiatives can be explained partly because 1992 was a presidential election year and President Bush was vulnerable to attack because of slow economic growth and a huge governmental deficit. (He was in fact defeated for reelection in large part because of poor economic conditions in the United States.)[1]

The conference made the term "sustainable development" known

139

throughout the world. The term means that present economic growth should not take place in such a manner that it reduces the ability of future generations to live well. Economic growth and efforts to improve the living standards of the few or the many should be sustainable, in other words, it should be able to be continued without undermining the conditions that permit life on earth, thus making future development impossible or much more difficult. The term represents an effort to tie economic growth and protection of the environment together, a recognition that future economic growth is possible only if the basic systems that make life possible on earth are not harmed. It also implies a recognition that economics and the environment are *both* important, that economic development and the reduction of poverty are essential to the protection of the environment. Sustainable development was endorsed by the conference and a new organization – the Sustainable Development Commission – was set up under the United Nations to monitor the progress nations are making to achieve it.

How does one evaluate this last world environmental conference? A pessimist would point to the fact that previous conferences announced fine goals also, which were endorsed by many nations, but few nations made significant efforts to achieve them. The pessimist would point also to the fact that on the urging of the Vatican and some conservative Islamic nations, most references to the importance of controlling population growth were dropped from conference documents. And the pessimist would point to the fact that neither the UN nor any other organization has any real power to make sure nations carry out the promises they made in the treaties they signed – and which must be ratified by the legislatures in many countries before they come into force – or in the conference declaration. The pessimist would call the 800-page document that presents ways the principles in the declaration can be carried out – titled Agenda 21 – as nothing better than an environmentalist's "wish list."

An optimist would point to the fact that this conference was attended by an unprecedented number of world leaders and gained unprecedented attention around the world. The optimist would remark that the wide acceptance of the goal of sustainable development was a significant accomplishment and will enable future efforts to promote economic growth to be evaluated in a way they never were before – and in both the rich and poor nations. The optimist would say also that many treaties made in the past have been obeyed by most of the nations signing them, not because there were provisions in them to force obedience but because the nations recognized that it was in their best interests to follow the treaty. The optimist would say that sustainable development is now recognized by many as being in a nation's best interests. While there will always be nations and individuals who care little about future generations, the fact that the principle of sustainable development has been publicly endorsed means that any efforts contrary to it are open to attack by peers and public opinion.

A judgment about the significance of the Rio conference will have to wait until enough time passes so the efforts or lack of efforts to carry out its principles can be seen. The statements by both the pessimists and the optimists are correct, and only the future will show which one has judged the conference's importance the best.

In this chapter we will examine some of the effects development has had on the air, water, and land of our planet. We will then focus on some of the dangers that have been created in the workplace and in the home. After looking briefly at the use of natural resources in the world, we will learn why the extinction of species is accelerating. Development's role in the extinction of human cultures will be explored also. The chapter will end with an explanation of what makes environmental politics so controversial.

The Air

Smog

Industrialization has brought dirtier air to all parts of the earth. From factories and transportation systems, with their telltale smokestacks and exhaust pipes, toxic fumes are being emitted constantly into the air. A few spectacular instances in the twentieth century resulted in large numbers of people becoming ill or dying because of the toxic gases in the air they breathed: 6,000 became ill and 60 died in the Meuse Valley in Belgium in 1930; 6,000 became ill and 20 died in Donora, Pennsylvania, in 1948; and in London tens of thousands became ill and 4,000 died in 1952.[2] (It was this last-mentioned instance that led the United Kingdom to pass various laws to clean up the air which have proved to be quite successful. By the late 1970s, 80 percent more sunshine reached London than had in 1952.)[3]

A number of industrialized countries, including the United States, have made significant progress in reducing air pollution in their large urban areas. Since the 1970 Clean Air Act was passed in the United States, lead was reduced by 95 percent, sulfur dioxide by about 30 percent, and particulates (tiny particles in the air) by about 60 percent. [4] And there is even evidence that a sharp decline in air pollution in the Arctic occurred in the 1980s, probably because the Soviet Union had shifted from coal and oil to the cleaner natural gas and Western Europe had reduced its sulfur dioxide emissions in order to combat acid rain.[5]

In spite of this progress, much remains to be done. In the early 1990s nearly 90 million people in the United States still lived in urban areas where the the air was considered poor enough to be called unhealthy during some days of the year.[6] In the late 1980s the US city with the worst air, Los Angeles, had unsafe levels of ground-level ozone, the chief factor in causing smog, during more than one-third of the days of the year and unsafe levels of carbon monoxide during about one-fourth of the year.[7] And even some of the progress cited above, the reduction of particulates, turned out to be less impressive than first thought. Studies in the early 1990s indicated that extremely small particulates, which were not illegal to release, were estimated to be causing up to 60,000 deaths a year in the United States.[8]

Third World countries that were starting to industrialize and modernize were facing air pollution problems even greater than those of the West. Third World cities such as Calcutta, Bangkok, Beijing, New Delhi, and Tehran had serious air pollution problems, caused by their increasing numbers of automobiles and factories. The World Bank reported that in the mid-1980s about 1.3 billion people, mainly in the developing world, lived in urban areas with unsafe levels of air pollution. The World Bank also reported that mainly in the rural areas of the poorer countries as many as 700 million people – mostly women and children – were being exposed to unsafe levels of indoor air pollution. This pollution was caused primarily by smoke from the wood, straw, and dung that were used as fuel for cooking and heating.[9]

Vehicles, such as this truck/bus, produce a lot of air pollution in the cities of the developing countries. (*Ab Abercrombie*)

The World's Dirtiest Air

Mexico City has earned the reputation as having the worst air in the world. The city has spread from about 50 square miles in 1940 to about 500 square miles in 1990. The density of population is twice that of the New York metropolitan area and four times that of London. Nearly 20 percent of the nation's people live in the city. In the early 1990s the ozone levels in the city were exceeding those set by the World Health Organization on 320 of the 365 days of the year. Just breathing the air was said to be the equivalent of smoking about two packs of cigarettes a day. The smog was so thick on some days that cars had to use their lights during the day, factories and schools were ordered closed, and the use of cars was restricted. About 70 percent of the air pollution comes from the 2.5 million cars, 200,000 buses, 35,000 taxis, and thousands of trucks in the city.

Source: Pete Hamill, "Where the Air Was Clear," *Audubon* 95 (January-February 1993), pp. 38–49.

With the collapse of communist regimes in Eastern Europe and in the Soviet Union in the late 1980s and early 1990s came evidence of startling amounts of pollution in those countries, pollution that had been kept secret for many years so it could not be used to criticize the regimes. Poland had some of the worst air pollution, in part caused by the heavy use of coal in its industry. By 1990 the air was so polluted in some industrial regions of Poland that adults and children suffering from respiratory problems were placed in salt mines, 600 feet below the surface, so they could breathe unpolluted air. The Polish Academy of Science reported that one-third of the nation was living in "areas of ecological disaster." [10]

Airborne Lead

The story of airborne lead illustrates well the connection between industrialization and air pollution. Scientists are able to estimate the amount of lead there was in the world's air in the past by taking core samples of ice in the Greenland ice cap. The ice, which represents past rainfall, shows that from 800 BC to the beginning of the Industrial Revolution around 1750, the amount of lead in the air was low. There was a major increase after 1750 and a huge increase after World War II when the use of leaded gasoline rose sharply. In 1965 the lead concentration in the Greenland ice was 400 times higher than the level in 800 BC. Other studies show that the bones of today's Americans contain 500 times more lead than those of prehistoric humans.[11]

Children are the ones who are most susceptible to harm from breathing lead. They inhale two to three times as much lead in the air per unit of body weight as do adults because their metabolic rates are higher and they have greater physical activity than adults. There is no known safe level of lead in the human body. High levels of lead poisoning can lead to death, but even low levels can cause learning difficulties and behavioral problems. In 1990 about 8 million children in the United States had dangerous levels of lead in their blood,[12] and black children were more likely to have high levels of lead than were white children.[13] Many of the black children live in old houses or apartments in the inner cities where they consume lead from lead-based paint flaking off from walls (which young children, who tend to put everything in their mouths, wind up ingesting), and from old lead water pipes.

There has been a significant improvement in reducing the amount of lead in the blood of Americans. Between 1976 and 1991 the amount dropped by nearly 80 percent.[14] Because of tighter federal government air pollution requirements, new cars were required to use nonleaded gasoline, and many experts believe that the reduced use of leaded gasoline is probably the cause of the lower lead levels in blood. In 1985 the US Environmental Protection Agency instructed oil companies to reduce the amount of lead in gasoline by 90 percent by the end of the year, and a total ban on leaded gasoline will go into effect in 1995. Most Western European countries are also restricting the use of lead in gasoline.

Another indication that the efforts to reduce the lead used in gasoline were having a beneficial effect can be seen by a study of the Greenland ice cap conducted in the early 1990s. This study of the lead concentrations in Greenland snow showed a drop in the lead concentrations to about the levels existing in the early 1900s, before the widespread use of leaded gasoline.[15]

Many developing countries have yet to come to grips with this problem. It has been estimated that children in Bangkok, Thailand, lose about 4 points on a test of intelligence by the age of seven because of lead poisoning. In the Mexico City metropolitan area, where in the early 1990s about 95 percent of the gasoline was still leaded, it was estimated that about 30 percent of the children had unhealthy levels of lead in their blood.[16]

Acid Rain

When fossil fuels are burned, sulfur dioxide and oxides of nitrogen are released into the air. As these gases react with moisture and oxygen in the atmosphere in the presence of sunlight, the sulfur dioxide becomes sulfuric acid (the same substance used in car batteries) and the oxides of nitrogen become nitric acid. These acids then return to earth in rain, snow, hail, or fog. When they do, they can kill fish in lakes and streams, dissolve limestone statues and gravestones, corrode metal, and possibly kill certain trees and reduce the growth of some crops. The effects of acid rain on human health are not yet known. Some scientists fear that acid rain could help dissolve toxic metals in water pipes and in the soil, releasing these metals into human water supplies.

In the United States, acid rain comes mainly from sulfur dioxide produced by coal-burning electricity-generating power plants in the Midwest and from the nitrogen oxides from auto and truck exhausts. It has caused lakes in the northeastern part of the country to become so acidic that fish and some other forms of life are unable to live in them. Other areas of the country, such as large parts of the South, Northwest, Rocky Mountains, and the northern Midwest, are especially sensitive to acid rain since the land and lakes in these areas contain a low amount of lime. Lime tends to neutralize the falling acid. An international dispute was created between Canada and the United States because a large amount of the acid rain falling on huge sections of Canada comes from industrial emissions in the United States.

Europe is facing a similar problem. Many lakes in Norway and Sweden are now so acidic that fish cannot live in them and about one-third of the forests in Germany are sick and dying. Much of the acid rain falling in northern and central Europe comes from industry in Britain, Germany, and France. The section of Europe with the greatest damage from acid rain lies in Eastern Europe. The efforts of the communist governments in that region to keep up with the West led to industrial growth fueled with lignite coal, which is cheap and abundant in the region but also extremely polluting. In one area where East Germany, Czechoslovakia, and Poland met more than 300,000 acres of forests have disappeared and the ground is poisoned by the huge amount of acid rain that fell there from the coal-fed power plants and numerous steel and chemical plants. Local foresters have dubbed the area the "Bermuda Triangle of pollution," as winds carry the sulfur dioxide and other pollutants to other areas of Europe.[17]

Acid rain was first observed in industrial England in the late 1800s, but nothing was done about it. In the 1950s the response to the increasing air pollution in the United States and Europe was to build tall smokestacks on factories so that emissions of toxic gases would be dispersed by the air currents in the atmosphere. These tall smokestacks led to a noticeable improvement in the air around many factories, smelters, power plants, and refineries, but the dispersal of noxious gases in the atmosphere gave more time for these gases to form into acid rain. We now realize that the tall smokestacks violated a fundamental law of ecology, one that biologist Barry Commoner has labeled the "everything must go somewhere" law.[18] Matter is indestructible, and there are no "wastes" in nature. What is excreted by one organism as waste is absorbed by another as food. When the food is toxic, the organism dies. Thus is explained the beautiful clear water in lakes with a high acid content: many forms of plankton, insects, and plants have ceased to exist there.

A 1983 study by the National Academy of Sciences concluded that acid rain could be curbed if sulfur dioxide emissions were reduced from coal-burning power plants and factories.[19] In 1984 nine European countries and Canada agreed to reduce sulfur dioxide emissions by at least 30 percent in the next decade. The United States, Britain, East Germany, Poland, and Czechoslovakia refused to join in this agreement. The

145

position of the Reagan administration in the United States at that time was that further scientific study was needed before costly action was taken. In 1986 the United States promised Canada that it would undertake a $5 billion research effort over the next five years to develop cleaner coal-burning technology.

In 1990 the US Congress passed the Clean Air Act, which calls for a large reduction of sulfur dioxide emissions from power plants. An innovative provision was put into the law that allows polluters to buy and sell their rights to pollute (the total amount of emissions indicated in the law must not be exceeded and the level will be lowered over time). The hope is that this provision will encourage polluters to find the cheapest way to cut their pollution.

In the same year the Clean Air Act was passed, a ten-year study of acid rain in the United States was completed. The authors of this most extensive study of the subject found that acid rain was a concern but not a crisis, and they admitted that many unknowns about the effects of acid rain still existed. The conclusion was that widespread damage to lakes and forests had not occurred in the United States because of acid rain.[20]

Somewhat similar findings about the forests in Europe came in a limited Finnish study covering the period 1971 to 1990. The study found that most forests continued to grow in spite of the air pollution. The authors concluded that although major damage to the forests of Europe had not occurred, the future was uncertain since airborne pollutants had changed the chemistry of the soil in ways that could have adverse effects over the long term. The scientists conducting the study calculated that about 3,000 square miles of forests in Europe, or about one-half of one percent of the Continent's total, had been killed by pollution.[21]

In the early 1990s, Japan and South Korea were receiving acid rain from the numerous coal-fed power plants in China. A concern was expressed in Japan and Korea over the plans by China to double its coal-burning power plants within a decade.

Ozone Depletion

The ozone layer in the atmosphere protects the earth from harmful ultraviolet rays from the sun. Scientists believe that life on earth did not evolve until the ozone layer was established. That layer is now being reduced by substances produced by humans, mainly in the developed nations. Chlorofluorocarbons (CFCs) – used as a propellant in aerosol spray cans, as a fluid in refrigerators and air conditioners, as an industrial solvent , and in the production of insulating foams – can destroy ozone. Ozone can also be destroyed by halons, which are chemicals used in fire extinguishers, and when nuclear bombs are exploded.

Scientists are agreed that any major depletion of the ozone layer would cause serious harm to humans, other mammals, plants, birds, insects, and some sea life. Skin cancer would increase, as would eye cataracts. Increased ultraviolet light could also adversely affect the immune system of humans, which protects them from many possible illnesses. As mentioned in chapter 6, one of the most harmful effects of a nuclear war would be the damage it would do to the ozone layer, which would affect life far beyond the combat area.

By analyzing past data, British scientists in the mid-1980s discovered that, during two months of the year, a hole was occurring in the ozone layer over the South Pole. Almost every year since it was discovered, the hole has continued to get larger. The hole (which is actually a significant reduction in the ozone normally found above that region, not a 100 percent decrease) galvanized the world to act to reduce the danger. About 60 nations met in Montreal, Canada, in 1987, and agreed to cut the production of CFCs by 50 percent by 1998. But further evidence that the depletion of the ozone layer was progressing faster than expected led the nations of the world to meet again – this time in London in 1990. The 90 nations attending that meeting agreed to speed up the phasing out of ozone-destroying chemicals. They agreed to halt the production of CFCs and halons by the year 2000. Less developed nations were given until 2010 to end their production and a fund was set up, mainly contributed to by the industrial nations, to help the poorer nations obtain substitutes for ozone-depleting chemicals.

New disturbing evidence of the ozone depletion danger was made public in the early 1990s. The US Environmental Protection Agency announced in 1991 that data from satellites, which had been collected over the previous 11 years, revealed that the ozone layer over large parts of the globe, including the layer above the United States and Europe, had been depleted by about 5 percent. This loss was occurring twice as fast as scientists had predicted. Based on the new findings, the agency calculated that over the next 50 years about 12 million people in the United States will develop skin cancer and more than 200,000 of them will die from it.[22]

Based on the new US evidence and on new data collected by an international team of scientists, which showed that the depletion was occurring in the dangerous summer months as well as in the winter,[23] 90 nations met in Copenhagen, Denmark, in 1992 and agreed to further accelerate the ending of ozone-destroying chemicals. All production of CFCs was to end by 1996 and halon production was to end by 1994. (Developing nations were again given a ten-year grace period to phase out the production of these two chemicals.)

Chlorine compounds enter the atmosphere mainly as a component of CFCs and it is chlorine that scientists now believe is causing the destruction of the ozone layer. One atom of chlorine can destroy 100,000 atoms of ozone. And since CFCs will remain in the atmosphere from 50 to 100 years, the destruction of the ozone layer will get worse. In the early 1990s scientists predicted that atmospheric chlorine would peak during the first decade of the next century and decrease thereafter.[24] Ozone is being made naturally all the time, so the situation will eventually improve. It has been predicted that, if the nations that have signed the ozone treaties follow through with their commitments, the ozone layer should regain full strength after about a century.[25]

The Depletion of the Ozone Layer: How to Protect Yourself

The American Cancer Society recommends that people try to keep out of the sun during the middle of the day – when ultraviolet rays are their strongest – and that they use hats and sun screen lotion when exposed to the sun. In New Zealand there is a public education campaign designed to end the fashion of getting a suntan.

In Chile people have taken to wearing sunglasses that block harmful ultraviolet light. All of these measures are sensible and will no doubt become more common as people's awareness of this new danger increases.

Carbon Dioxide

The release of carbon dioxide into the atmosphere from the burning of fossil fuels may be causing a warming of the earth's climate. This warming, called the greenhouse effect, has been discussed in chapter 4. No affordable technology exists at present to prevent the release of carbon dioxide when fossil fuels are burned.

The Water

Development, to date, has tended to turn clean water into dirty water as often as it has turned fresh air into dirty air. In the United States the deterioration of the nation's rivers was dramatized in the late 1960s when the Cuyahoga River, which flows through Cleveland, caught fire because it was so polluted. That event helped prod the US Congress into passing the Clean Water Act of 1972, which set a ten-year goal to return the nation's waterways to a state where they would be "fishable, and swimmable." Ten years later, many US rivers, streams, and lakes were cleaner than they had been when the Act was passed, but many still remained too polluted to allow safe fishing or swimming. This fact was illustrated by the recommendation that the state of New York's Department of Health made to that state's residents in 1983. The department recommended that they should eat no more than one meal a month of freshwater fish caught in the state and that pregnant women, women of child-bearing age, nursing mothers, and children under 15 should eat no fish at all from the state![26]

By 1990 the $75 billion that had been spent in the United States on upgrading sewage treatment facilities during the previous two decades had resulted in a significant improvement of the nation's waters. A survey revealed that 80 percent of the nation's rivers and streams were now safe for fishing and 75 percent were safe for swimming. But that survey indicated also that about 130,000 miles of rivers were still unsafe for fishing and 150,000 miles were unsafe for swimming.[27]

Why was there still a significant problem after this large expenditure and 20 years of effort? A large part of the reason was that little progress had been made in reducing the pollution from urban and agricultural runoffs. Especially during storms, huge amounts of polluted water from city streets and the lawns of houses drain directly into rivers and lakes, untreated by local sewage treatment plants, and huge amounts of water drain from farms, water laden with pesticides, herbicides, and fertilizers.

Water pollution in the USA is partly caused by large amounts of pesticides, herbicides, and fertilizers which run off from fields during storms. (*Vince Winkel*)

Other developed countries are also experiencing serious water pollution problems. As Lester Brown has nicely put it, if Johann Strauss were to write his famous waltz today he would have to call it the "Brown Danube." The Danube, which flows through central Europe, was so badly polluted in the late 1970s that it was illegal to swim in it, and the Rhine, even after billions had been spent to clean it up, was still considered the world's dirtiest river.[28] The former Soviet republics, and Eastern European countries that had been part of the Soviet empire, have some of the most severe environmental problems of any developed country, including serious water pollution problems. Consider the following: In the late 1980s only 40 percent of the wastewater in Czechoslovakia was adequately treated. Half of the cities in Poland, including Warsaw, and 35 percent of the industries did not treat their wastes at all. And in the Soviet Union many large cities had no sewage treatment facilities. Half of the drinking water in Czechoslovakia failed to meet national health standards.[29] Half of the drinking water in Russia in the early 1990s was contaminated.[30] One exception to the rather gloomy picture in Europe is the River Thames, which flows through London. Once known as being a very filthy river, it has been cleaned up to such an extent that fish have now returned to it.

What is causing the polluted water in the developed nations? Industry must take a large part of the blame since traditionally industrial wastes have been dumped into nearby water as often as they have into the air overhead. Many industries are no longer dumping wastes into nearby rivers, but some dumping still goes on. The source of much of the most serious water pollution today in the developed countries is chemicals. The chemical industry has had a huge growth in the industrial world since World War II. Chemicals are now finding their way into waterways, many of which are being used for drinking water.

Pesticides are one example of dangerous chemicals. In the late 1980s the US Environmental Protection Agency (EPA) found at least 46 pesticides in the groundwater of 26 states in the United States.[31] A national survey by the EPA, published in 1990, found pesticides in about 10 percent of all community water systems. The agency estimated that about 1 percent of these systems had potentially unsafe concentrations of pesticides.[32] In some regions of the country the contamination was much higher than the national average. Surveys showed that as many as 20 to 30 percent of the community wells in Minnesota and Iowa and 30 to 60 percent of the private wells in those states may contain pesticide residues.[33] It has been estimated that 90 percent of the pesticides that are applied to crops and lawns end up as runoff in waterways.[34] And the standard water treatment facility is usually unable to remove pesticide residues.

In addition to the problem of the unsafe disposal of wastes by the chemical industry, which will be discussed in the next section of this chapter, many of the 50,000 different chemicals that the US chemical industry is producing are not now being disposed of by consumers in a manner that avoids contaminating water supplies. An especially dangerous fact is the discovery that about one-half of the underground water upon which about 100 million Americans depend for all or some of their drinking water is threatened with chemical contamination or is already contaminated.[35] These underground aquifers, some of which took millions of years to form, were once thought to be safe from pollution, but it is now recognized they are threatened. Once these waters are contaminated it is extremely difficult, and many times impossible, to remove the contamination. The groundwater aquifers are replenished slowly from the land above. When that land is contaminated by suburban cesspools, industrial wastes, and chemicals used on farms, the groundwater becomes polluted. The experience of communities on Long Island in New York was that, after the land above the aquifer was urbanized, it took about 20 years for the aquifers, from which many of these communities got their drinking water, to become contaminated.[36]

The US Government's *Global 2000 Report* projects that the demands for freshwater in the world will increase 200 to 300 percent from 1975 to the year 2000. Population growth alone will lead to a doubling in the demands for freshwater in half the countries of the world.[37] The largest user of water is agriculture, which takes 70 to 80 percent of that used by humans. Agribusiness uses a huge amount of water for irrigation, and its demands are expected to grow as industrial farming techniques continue to spread in the United States and the Green Revolution spreads around the world. Except for regional and temporary shortages, the United States has never had a shortage of freshwater; indeed Americans are as profligate in their use of water as they used to be of oil before the oil shocks. The situation may change as heavy demands on freshwater by agriculture occur. The continued development of states like Nevada and Arizona has led to such extensive pumping of groundwater that the land is sinking (4 feet in Las Vegas in only 20 years) and huge cracks are opening up in the land (some near Phoenix). In the Third World, widespread deforestation will make water supplies more erratic and growing populations will undoubtedly lead to new conflicts over the availability of water, such as those that have taken place in the past between India and Pakistan, Israel and Syria, and Mexico and the United States.

The Land ●

Whenever development has occurred, its effect on the land has been profound. The economic growth that comes with development increases the amount of goods and services available for human consumption. More natural resources from the land are required for the production of these goods, of course, and their extraction disturbs the land greatly. But even more widespread are the changes to the land that come with the disposal of the goods after they are no longer of use, and of the wastes that are created in the manufacture of the goods. Many of these wastes are artificial substances that never existed before in nature; thus nature has few, if any, ways of breaking them down into harmless substances. Development also affects the vegetation on the land, in some ways reducing it and in some ways helping to preserve it. In this section we will focus on two of the many changes to the land that come with development: the disposal of wastes and deforestation. These two changes are affecting many human beings in such direct ways today that it is important that we look at them closely.

Solid Wastes

It seems to be a common occurrence in a number of developed countries that, as more goods and services become available, more are desired and less value is placed on those already in hand. After the end of World War II, an unprecedented period of economic growth in the industrialized world took place, leading to a huge increase in the consumption of material goods.

As consumption rose so did wastes. "Throw-away" products that were used briefly and then discarded became common as did items that wore out quickly. Such facts disturbed few people in the United States since they found enjoyment in buying new, "better" products. Many such products were relatively cheap in the 1950s, 1960s, and early 1970s since energy and other raw materials were inexpensive. Between 1970 and 1988 the annual amount of solid wastes generated in the United States grew by about 25 percent, until it reached about 1,500 pounds per person.[38]

In the early 1990s the EPA published new regulations requiring all cities and towns in the United States to improve their landfills or dumps where their solid wastes were being disposed of, so pollutants would not leak from them into the groundwater or into nearby waterways. The municipalities had two years to comply with the regulations. The EPA estimated that of the 6,000 landfills in the country, about one-half of them would have to close because the cost of meeting the new regulations would be too much for many towns to bear.

One obvious way cities could respond to the new requirements was to get their citizens to produce fewer solid wastes and to start recycling trash, which many communities did to reduce the amount of trash going to their landfills. Another way to reduce trash was to make citizens pay variable costs for the disposal of their solid wastes, based on amount and type. Seattle, Washington, is an example of a US city that has successfully followed that principle. Seattle began charging its citizens according to the amount of trash they put out for disposal. Yard wastes, such as grass clippings, if separated by the citizens so the city could use them for composting, were charged at a much lower rate than regular trash, and paper, glass, and metal (which could be recycled) were hauled away free. (One-half of all trash in the United States is paper and paperboard, and about 30 percent is yard waste.) Seattle, which was already more environmentally conscious than most other American cities, found that in the first year it started charging its citizens for the amount of waste they produced, the total tonnage the city needed to haul to the landfills fell by about 20 percent. The amount of wastes being recycled by its citizens rose from about 25 percent to over 35 percent.[39]

Besides creating ugliness throughout the land, the wastes soon brought other changes, changes to the health of people who inadvertently came in contact with them. Toxic wastes became the environmental problem of the 1980s in the United States, as thousands found that their health was, or was possibly, being affected by chemical wastes. These wastes came after World War II from the manufacture of many new products from chemicals – everything from new fabrics, deodorants, toothpastes, and plastic containers to new drugs, pesticides, and fertilizers.

Toxic Wastes

The first warning of the danger of toxic wastes came from Japan. In the 1950s and 1960s hundreds of people were paralyzed, crippled, or killed from eating fish contaminated with mercury that had been discharged into Minamata Bay by a chemical plant. In the late 1970s the warning came to the United States. Many people in a residential district of Niagara Falls, New York, were exposed to a dangerous mixture of chemicals that were seeping into their swimming pools and basements. Most of these people did not know, when they bought their homes, that the Hooker Chemical Company had dumped over 20,000 tons of chemical wastes in the 1940s and 1950s into a nearby abandoned canal, ironically known as Love Canal. News of the Love Canal disaster spread through the country as the story of the contamination slowly came out in spite of the denials of the chemical company and the apathy of the local government.[40] Eventually hundreds of people were evacuated from the area. The state and federal governments bought over 600 of the contaminated homes. After putting a "wall" of clay and plastic around the buried toxic waste, the federal government later declared much of the Love Canal neighborhood fit for resettlement. The name of the neighborhood was changed from Love Canal to Black Creek Village.

The EPA estimates that US industry produces about 1 trillion pounds of new toxic wastes each year.[41] Since 1989 the EPA has been requiring some of the producers of toxic wastes to report annually how much waste they are producing. Two industries that produce the largest amount of toxic wastes – the mining and oil-exploration industries – are exempt from the reporting requirement. The remaining industries that produce toxic wastes reported in 1991 that they produced 38 billion pounds of chemical wastes. Half of those wastes were produced by the chemical industry, with other significant amounts produced by the metal, oil refining, paper, and plastics industries. This total was 30 percent less than had been produced in 1988. Of the 38 billion pounds produced in 1991, about one-half of it was recycled. The head of the EPA, in the following statement, tried to help people understand how much 38 billion pounds is : "Thirty eight billion pounds is the equivalent of a line of tank trucks that stretches halfway around the world."[42]

Governmental and Industrial Responses to the Waste Problem

In 1980 the US government created a $1.6 billion fund to finance the cleaning up of the worst toxic waste sites. The law which set up this fund (popularly called Superfund) allowed the government to recover the cost of the clean-up from the companies that dumped wastes at the sites. In 1986 $9 billion more for the clean-up was approved by the US government, to come mainly from a broadly based tax on industry and a tax on crude oil. The Congressional Office of Technology Assessment has estimated that it will require about 50 years and $100 billion to clean up toxic waste dumps in the country.[43] By the early 1990s the EPA had identified nearly 40,000 toxic waste sites in the country as being potentially hazardous. Of the 1,200 sites that were identified as the worst and needing priority action to clean them up, only about 150 had been cleaned up by 1992.[44]

There are other ways government can help control the waste problem. Barbara Ward, the late British economist, mentions four ways a government can encourage the reduction of wastes and promote the reuse of wastes: 1) it can make manufacturers pay a tax that could cover the cost of handling the eventual disposal of their products; (2) it can stimulate the market for recycled products by purchasing recycled products for some of its own needs; (3) it can give grants and other incentives to cities and industries to help them install equipment that recycles wastes; and (4) it can prohibit the production of nonreturnable containers in some instances.[45] The last-mentioned device has been used by nine US states. They have effectively banned nonreturnable (throw-away) beer and soft drink containers, and are getting about 90 percent of the refillable containers recycled.[46]

Inefficient and wasteful technologies and processes to produce goods are still common in the United States and other developed nations, since many of these were adopted when energy was cheap, water plentiful, many raw materials inexpensive, and the disposal of wastes easy. Some industries now realize that they can increase their profits by making their procedures more efficient and producing less waste. One such company is 3M, which according to one study, has reduced its pollution as well as increased its profits, "not by installing pollution control plants but by reformulating products, redesigning equipment, modifying processes...[and] recovering materials for reuse."[47]

Deforestation

In the most authoritative study of deforestation ever conducted, the UN's Food and Agriculture Organization (FAO) reported in 1991 that the destruction of tropical forests was accelerating. Using satellite photographs and spot checks on the ground by FAO workers and their consultants, FAO estimated that about 40 million acres of forests were cut down in 1990, a 40-percent increase over amounts being cut a decade earlier. The amount being cut down in 1990 equaled the size of the state of Washington, in the United States, or about about one acre every second. [48] In most poor countries that are undergoing a large growth of population and still have sizable forests, deforestation has been accelerating rapidly. It has been estimated that the developing nations lost about half their forest cover in the twentieth century.[49]

The cutting down of tropical forests is accelerating. (*USAID Photo Agency for International Development*)

In contrast to the situation in the developing countries where the tropical forests are located, the forests in the developed nations actually increased in the twentieth century. It is estimated that as marginal farm land was taken out of production and trees were allowed to return to the land, forests grew in the industrialized countries from about 7.7 million square miles in 1900 to about 8 million square miles in 1985.[50]

Deforestation is a serious problem because it can lead to erosion of the land, it can cause the soil to harden, and it can make the supply of freshwater erratic. Scientific studies support the hypothesis that deforestation can lead to significant changes in the climate. These changes usually mean less rainfall.[51] Sometimes deforestation leads to too much water in the wrong places. Serious floods are occurring now in India in areas that had never experienced flooding; it is believed that the cutting of forests in the Himalayan Mountains, the watershed for many rivers in India, is causing the flooding. Rioting has even been reported among some of the tribal peoples of India who are protesting the cutting of their forests by commercial firms.

Much of the destruction of the tropical rain forests, however, is carried out not by commercial logging firms, although they do a significant amount of cutting, but by small-scale farmers.[52] Other forest destruction has been caused by colonization projects and the creation of pastures for cattle raising. In the early 1970s the Brazilian government began a large colonization project in the Amazon basin, moving people in from the poverty-ridden northeastern section of the country. It was hoped that the resettlements would help reduce the poverty in the northeast and provide food for an expanding population. Unfortunately, both hopes faded as colony after colony failed. The main reason for the failure was that tropical forest land is actually not very fertile, in spite of the huge trees growing on it. Such trees get their needed nutrients directly from decaying leaves and wood on the forest floor, not from the topsoil, which, in many places is thin and of poor quality. This fact explains why many of the settlers had experiences similar to that of the following Brazilian peasant who described what happened to his new farm in the Amazon:

> The bananas were two feet long the first year. They were one foot long the second year. And six inches long the third year. The fourth year? No bananas.[53]

Small-scale farmers clearing land for crops are one of the main causes of deforestation, as in this Brazilian rain forest. (*Campbell Plowden/Greenpeace*)

Some scientists now believe that the mysterious decline of several great civilizations in the past, such as the Khmer civilization in Southeast Asia and the Mayan civilization in Central America, was caused by

156

over-intensive agriculture on basically fragile land. And recent research supports the theory that the total deforestation of Easter Island, where a former civilization carved over 600 huge stone heads, was the direct cause of the collapse of that civilization since it led to the loss of fertile soils and fuel.[54]

The cutting of the trees in a tropical forest puts a severe strain on the soil since the trees protect the soil from the violent rains that are common in the tropics. And once the soil is washed away, it is not easily recreated. Some studies now estimate that from 100 to 1,000 years are needed for a mature tropical forest to return after human disturbances have taken place.[55]

If only small plots of the forest are cleared, regeneration of the forest is possible. Some peoples have practiced what is known as shifting cultivation in the tropical forests. They clear a piece of land and farm it for a year or two before moving on to a new piece of land. As long as this remains small-scale, the damage to the forest is limited, but any large-scale use of this type of agriculture can lead to irreversible damage to the forest.

Some tropical soils contain a layer known as laterite, which is rich in iron. When these soils are kept moist under a forest they remain soft, but if allowed to dry out, which happens when the forest cover is removed, they become irreversibly hard – so hard that they are sometimes used for making bricks.

There is a danger that the destruction of forests will contribute to global warming. Trees that are burned after they are cut, which is common when the forest land is cleared for settlements or for farming, release carbon dioxide into the atmosphere. The great tropical rain forests contain a huge reservoir of carbon and have been described by some as the "lungs" of the earth, absorbing carbon dioxide and releasing oxygen.

In Central America and in Brazil, large areas of forests are being cut down to make pastures for the raising of cattle. The cattle are being raised mainly to supply the fast-food hamburger market in the United States. The growing of cattle on large ranches for export, does not, of course, do anything to solve the food problems in the exporting countries, or to provide land to the landless.

A positive development that might help preserve tropical forests is the recent interest by Western pharmaceutical companies in plants from the forests from which they can make drugs to combat AIDs, cancer, heart disease, and other illnesses. One example of this is Merck & Company, the world's largest drug company, which is paying a research institute in Costa Rica to search for certain plants, microorganisms, and insects for possible medical use. The company has also agreed to share any resulting profits with Costa Rica.[56] Efforts such as this can help convince a developing nation that income can be earned from the forest without damage to it.

Recent research – although limited in scope – also suggests that more

157

income can be made by "mining" the forests for plants useful in traditional medicine (which is utilized by 80 percent of the world's people) and for products that can be marketed in the West (such as nuts and cosmetic oils), than by cutting the trees for lumber or by using the land for farming.[57] Local people can also earn income from what has been called "eco-tourism." This type of tourism focuses on the growing number of people from the developed nations who wish to visit tropical forests and other spots that have been left more or less in a natural state. If evidence exists that local people can earn more income by letting the forests remain than by cutting them down, a strong argument can then be made supporting their preservation. Also, local people can be enlisted in the efforts to prevent deforestation since they will have an economic stake in the preservation of the forests.

It is mainly the landless and the poor around the world today who are assaulting the remaining forests for agricultural land and for fuel. As poverty is the root cause of the hunger problem, discussed in chapter 3, and one of the root causes of the population explosion, discussed in chapter 2, so also is it at the root of the deforestation problem. Development can reduce poverty, and when it does this for the multitude, it can reduce the threat to the world's forests. Development can also lead to the destruction of the forests as they are cleared for cattle farms, for lumber, for commercial ventures, and for human settlements. As with the population problem, development in its early stages seems to worsen the situation; but development that benefits the many and not just the few can eventually help relieve it.

The Workplace and the Home ● ● ● ● ● ● ● ● ● ● ● ● ● ●

Cancer

Cancer is often considered to be a disease of developed countries. It is estimated that one out of four people in the United States alive at present will contract cancer and many of them will die from it.[58] Cancer now kills more children than any other disease, although accidents are still the number one cause of death of children. It is commonly believed by the general public that exposure of workers to cancer-causing substances – carcinogens – in the workplace, and the exposure of the general population to pollution in the air and water and to carcinogens in some of the food they eat, are the main causes of this dreaded disease. There is no question that many workers – such as the millions of people who worked with asbestos – have been exposed to high levels of dangerous substances. But scientists do not now believe that contamination at the workplace is the main cause of cancer; nor do they believe that air and water pollution or food additives are causing most of the cancer cases. The consensus among leading cancer experts at present is that smoking and diet, mainly a high-animal fat and low-fiber diet, are the main causes.[59] Some experts, however, fear that while chemicals cannot be blamed for most cancer today, there is a possibility that chemical-related cancers may increase greatly in the future because of the large increase in the production of carcinogenic chemicals since the 1960s. Cancer can occur 15 to 40 years after the initial exposure to a carcinogen, so chemicals may yet prove to be a major culprit.

Pesticides

The story of pesticide use illustrates well the dangers that new substances, which have become so important to modern agriculture, have brought to people at their workplace as well as in their homes at mealtime. Rachel Carson is credited with making a whole nation aware of the dangers of persistent pesticides such as DDT. Her book, *Silent Spring*, which appeared in 1962, shows how toxic substances are concentrated as they go up the food chain, as big animals eat little animals. Since most of the toxic substances are not excreted by the fish or animal absorbing them, they accumulate and are passed on to the higher animal that eats them. Carson's warning led to a sharp reduction in the use of long-lived pesticides in many developed countries; but if she were alive today (she died of breast cancer in 1964), she would probably be disturbed to learn that short-lived, but highly toxic, pesticides are now increasing in use in the United States. The use of herbicides has especially increased dramatically as farmers, railroad companies, telephone companies, and others find it cheaper and easier to use these chemicals to get rid of unwanted vegetation than to use labor or machines. These new highly toxic pesticides pose a special risk to the workers who manufacture them and to the farmers who work with them in the fields. (They have probably caused the death of large numbers of fish in the southern United States as airplanes spraying crops with these so-called "soft" pesticides have also sprayed lakes, bays, and even some people.)[60]

Although DDT was banned for use in the United States in 1972, residues of it could still be found in most people in the United States 20 years later.[61] One study in 1993 found that women with the highest exposure to DDT had four times the risk of getting breast cancer as those with the least exposure.[62]

Pesticide use is increasing in the less developed nations – not just the use of short-lived pesticides but of persistent pesticides such as DDT as well. Of the pesticides produced in the United States for export, 25 percent are substances that are banned, highly restricted, or unregistered in the United States.[63] US law explicitly permits the sale of these substances to foreign nations, and US companies, as well as many in Europe, have increasingly turned to the overseas market to sell their products as more restrictions on the use of pesticides occur in the developed nations.

Pesticide Poisoning

The World Health Organization estimates that as many as 20,000 deaths occur annually around the world because of pesticide poisoning and 1 million people are made ill. Probably most of the people harmed by pesticides are in developing countries, where many farm laborers who are using the pesticides are illiterate and cannot understand the warnings on the pesticide containers. Many other workers, such as in the cotton fields in Central America, end up having themselves, their water supplies, and their homes, which are often near the fields, sprayed with pesticides from the air, some as many as 40 times during a growing season. In the late 1970s Guatemalan women living in cotton-growing areas had the highest concentration of DDT in their breast milk ever found in the Western world. Tropical fruits being grown for export are sprayed heavily to keep them free of blemishes. A vicious circle is created in which the heavy use of pesticides leads to the killing off of natural predators and the evolving of pests that are resistant to the pesticide, which in turn leads the farmer to turn to new and more toxic pesticides.

Sources: World Health Organization (WHO), *Public Health Impact of Pesticides Used In Agriculture* (Geneva, Switzerland: WHO, 1990), p. 86; and Martin Wolterding, "The Poisoning of Central America," *Sierra* (September/October 1981), p. 64.

Many of the agricultural products on which banned pesticides are being used are imported into the United States. The US Food and Drug Administration (FDA) estimated in the early 1980s that at least 10 percent of the food coming into the country had illegally high residues of pesticides on it.[64] And it is believed that the real percentage of imported contaminated food is much higher than the FDA estimates because the administration does not test for many contaminants. The FDA finds mysterious, unknown chemicals on the foods at times, which is not surprising, since chemical companies are permitted to export unregistered pesticides, pesticides that have never been tested by the government. Illegally high residues of pesticides have been found on imported meat, coffee, beans, peppers, and cabbages. Many of these products are sold in American supermarkets before the FDA tests on them are completed.

The US Department of Agriculture has estimated that 32 percent of US crops were lost to pests in 1945 whereas in 1984, despite a large increase in the use of pesticides, the loss due to pests had risen to 37 percent.[65] This situation may explain why a number of agricultural experts are now advocating a more balanced program for controlling pests. A selective use of pesticides would go along with the use of biological controls, such as natural predators, and other nonchemical means to control pests.

In the early 1990s the US government announced that it was going to try to reduce the amount of pesticides used on US farms. A five-year study by the US National Academy of Sciences on the effect of agricultural chemicals on children was published in 1993. It criticized the method the government had been using to calculate the safe amount of pesticide residue on foods. It found that the risk calculations by the government had not taken into account the fact that people are also exposed to pesticides from sources other than on foods, such as in their drinking water, on their lawns, and on golf courses. It found that infants and children might be especially sensitive to pesticide residues on food. They consume 60 times the amount of fruits adults do, according to their weight, so are getting higher doses of the pesticides that are used on fruits. And this is taking place early in their life. The head of the committee that prepared the report drew the following conclusion: "Pesticides applied in legal amounts on the farm, and present in legal amounts on food, can still lead to unsafe amounts."[66]

Pesticides have played an important role in the successes of the Green Revolution; it is doubtful that food production would have stayed ahead of population growth in the world without them. What seems to be called for now is a highly selective use of pesticides, not their banishment.

Artificial Substances

Development has led to the introduction of many artificial substances that have never been adequately tested to verify their safety. The National Research Council of the US National Academy of Sciences concluded, after a three-year study in the mid-1980s, that tens of thousands of important chemicals had never been fully tested for potential health hazards. The Council found that this included about 90 percent of the chemicals used in commerce, 60 percent of the ingredients in drugs, 65 percent of pesticide ingredients, 85 percent of cosmetic ingredients, and 80 percent of food additives.[67] The general conclusion of the Council was that we are ignorant of the potential harm that might be caused by many products we come in contact with in our lives today.

Not everything causes cancer, of course, but development has brought forth so many new products in such a short time that we cannot be sure which ones do and which do not. Barry Commoner shows that new products often bring large profits to the first industry that introduces them, so there is a strong incentive for industries to be innovative. New products, especially in the United States since World War II, are often made of synthetic materials that pollute the environment, but the pollution usually does not become evident until years after the introduction. Commoner states that "by the time the effects are known, the damage is done and the inertia of the heavy investment in a new productive technology makes a retreat extraordinarily difficult."[68]

The Use of Natural Resources

Since the world's population is growing exponentially, as we learned in chapter 2, it is probably not surprising that the consumption of nonfuel minerals is also growing exponentially. But, unlike petroleum, the supplies of minerals are not becoming exhausted. In fact, the known reserves of all major minerals actually increased between 1950 and

1980 as more deposits of ores were discovered.[69] Another great difference between nonfuel natural resources and energy supplies is that the actual cost of producing most minerals has decreased over the past century.[70] This reduced cost has occurred, even as lower-grade ores are being mined, because of advances in technology – such as better exploration techniques, bigger mechanical shovels to dig with, bigger trucks to haul the ore away, and bigger ships to transport it to processing plants. Whether new technology will continue to keep the cost of minerals low in the future is a subject that is debated by scientists and economists. As ores containing a lower concentration of the desired minerals are mined and less accessible deposits are turned to, processing costs will rise. Also, mineral extraction is highly energy-intensive and rising energy costs will directly affect the price of minerals. Some analysts have observed that mineral prices in the past did not reflect the true environmental costs of extracting and processing the minerals, but with new pollution laws in most industrial countries, the mining industry will have to assume more of these costs.

One trend that is apparent is that most industrialized nations are becoming more dependent on foreign countries for their minerals. The United States is a mineral-rich country; in the 1950s it was nearly self-sufficient in the most important industrial minerals. By the late 1970s it was self-sufficient in only seven of the 36 minerals essential to an industrial society.[71] Western Europe and Japan are even more dependent on imported minerals than is the United States. This increasing dependency on ores from foreign countries, many of which are essential for the advanced technologies common in the United States, has strongly influenced US policy toward the Third World, where many of the minerals are found.

There are four main steps a country can take to counteract shortages of a needed material if it cannot locate new rich deposits of the ore: (1) it can recycle waste products containing the desired material; (2) it can substitute more abundant or renewable resources for the scarce material; (3) it can turn to ore deposits having a lower concentration of the needed mineral; or (4) it can reduce its need for the material. All of these options have some negative features.

Recycling

It is generally agreed that more recycling of waste material needs to be done in the United States. In the late 1980s recycling became relatively popular in the country because more citizens became aware of environmental problems and because many towns were faced with trash dumps that were becoming filled. (New dumps were becoming very expensive to open because of tighter federal government regulations.) This new popularity of recycling led to more material being collected than could be sold. Large amounts of the recyclable material even ended up in Europe, where it caused difficulties for their recycling efforts. It flooded

European markets so that some communities could no longer find a buyer for the recyclable material they were collecting.[72]

A demand for recycled material will gradually grow in the United States as more industries start using it. In fact by the mid-1990s this was taking place as a growing economy emerged and new plants able to process recycled material began operating. One solution that has been suggested to promote recycling is that government should require the use of more recycled material, such as in newsprint. The US government did take this step in 1993 when President Clinton ordered all federal government agencies, including the military, to purchase paper with a minimum of 20 percent recycled fibers in it. But even with respect to newsprint, there is no simple solution since newspaper cannot be recycled indefinitely as the quality of the fibers degrades.[73]

Even with the new interest in recycling in the United States, the country is still not doing as much of it as other industrial nations do. In the early 1990s the United States was recycling about 15 percent of its trash [74] while Japan was recycling about 50 percent.[75] Western European countries also do much more than the United States. Germany passed a law in 1991 that requires much of its packaging to be collected and recycled. Also, German automakers have been required to redesign their cars so that they can be more easily dismantled for recycling when they wear out. The chairman of Volkswagen stated: "We must adopt the cyclical processes on which the whole of nature is based."[76] One of the central features of the German law is that although the changes are mandated by the government, the implementation of them is given to private companies motivated by an economic incentive .

While recycling is desirable, it is only a partial solution to resource shortages and to pollution by the minerals industry. Recycling also creates pollution and uses energy. Furthermore recycling is now only about 30 percent efficient for most used metals.[77] The move by the US soft drink and beer industry to use aluminum cans that can be recycled is obviously not the final solution to the litter problem. About 40 percent of aluminum cans are never collected for recycling,[78] and the manufacture of aluminum uses a lot of energy. Probably a better solution was the move by some American states, as mentioned above, to require returnable soft drink and beer containers to be used in their states instead of throwaways. Oregon's experience with its container law has been that highway litter was significantly reduced, recycling was stimulated, the price of beverages remained about the same, and new jobs were created.[79] In the states that have container laws about 80 percent of all containers are returned to the stores.[80]

An examination of the reaction by some companies to the efforts of a few states to require that deposits be collected on beverage containers to encourage their return for recycling or reuse illustrates the economic pressures that discourage real recycling. One investigator found that beverage companies (both soft drink and beer), companies that make glass beverage containers, and retail grocery chains financially support

163

the Keep America Beautiful campaign, which encourages recycling and the picking up of litter by citizens, largely the containers that these companies produce or sell. (Studies have shown that 50 percent of US litter is beverage containers.) The investigator found also that these companies were strong supporters of efforts to defeat attempts by states to pass legislation requiring returnable containers and deposits. A front organization funded by these companies and others spent $2 million in 1987 to defeat an attempt to pass a deposit law in the District of Columbia, the area around the nation's capital city.[81] These companies complain that having to reprocess empty containers will increase their expenses.

The "throwaway" economy that developed in the United States after World War II still exists. The new efforts to recycle are a step forward, but much remains to be done. While it is certainly good that citizens become involved in picking up litter along their highways, better yet would be a system that discouraged the litter. Recycling is much better than burying containers, but better yet would be to reuse them. The investigator who explored the hidden motives behind industries' support for efforts to reduce litter in the country commented in an interesting way on the situation in the United States :

> Only in America could custom compel the discarding of a perfectly good vessel simply because someone had quaffed the contents, but that's what we do with 50 billion cans and bottles every year. An additional 50 billion or so are "recycled," a uniquely American interpretation of the word because they too are discarded, then crushed, melted, and remade rather than simply washed and refilled. It's as if we were a nation of dukes and earls, pitching our brandy snifters at the hearth.[82]

Denmark has banned throwaways, but this action is unlikely to be taken in the United States any time soon.

Substitution

When a material becomes scarce, it is sometimes possible to substitute another material for it which is more abundant or to use a renewable resource in place of the scarce item. For example, the more abundant aluminum can be used in place of the scarcer copper for most electrical uses. Difficulties arise at times when the substituted material in turn becomes scarce. Plastic utensils and containers replaced glass products in most US kitchens because of certain advantages plastic has over glass, such as being less breakable and lighter in weight. But plastics are made from petrochemicals, which are now becoming scarcer. Also, the plastics industry produces more dangerous pollutants than does the glass industry. Another limitation to substitution is that some materials have unique qualities that no other materials have. Tungsten's high melting point, for example, is unmatched by any other metal. And substitutions can produce disruptions in the society, causing some industries to close and new ones to open. The last-mentioned point can mean, of course, new opportunities for some people and fewer for others. New ways of doing things can also be substituted for old ways, sometimes resulting in a reduced use of resources. The trend in some businesses to use communications in place of transportation (business meetings with participants on video screens instead of physically being present) might be such a development.

Mining of Low-Grade Ores

Many of the deposits with the highest concentration of the desired minerals have now been mined, but there are large, less rich deposits of many desired minerals scattered around the world. These can be, and in many places are, mined. There are significant costs incurred when such mining takes place. The cost of the mining increases, since more ore must be processed, mines must be bigger, more energy and water must be used. Because more ore must be processed, more wastes are produced. Large strip mines are often used and these have a devastating effect on the land. Even the best attempts to restore the strip-mined land are costly, and very imperfect.

Reducing Needs

The fourth way to counteract shortages of a material is to reduce the need for the material. Many consumer goods – such as automobiles and clothes – become obsolete in a few years as styles change. This planned obsolescence leads to a high use of resources. Many products also wear out quickly and must be replaced with new ones. More durable products could be designed by US industry, but they would often be more expensive. It is probably because of this reason that US industry generally does not make such products. Higher prices would mean fewer sales, a slower turnover of business inventories, and thus probably lower profits. They could also mean fewer jobs.

Overdevelopment

Perhaps a good way to end this section is to explain the concept of overdevelopment. According to the Australian biologist Charles Birch, "overdevelopment of any country starts when the citizens of that country consume resources and pollute the environment at a rate which is greater than the world could stand indefinitely if all the peoples of the world consumed resources at that rate."[83] From this perspective, it can be seen that the United States could be considered the most overdeveloped country in the world, followed closely by many other industrial countries. People in the United States, who constitute about 5 percent of

165

the world's population, consume about 30 percent of the world's annual use of natural resources, and do so, as this chapter has shown, with devastating effects on the environment. This devastation is being reduced as new environmental laws are enacted and gradually enforced in the developed world, but it has not been reduced to such an extent that the concept of overdevelopment is outdated.

The Extinction of Species ● ● ● ● ● ● ● ● ● ● ● ● ● ● ●

No one knows for sure how many species of living things there are on the earth. Biologists today generally make educated guesses that the number is between 10 million to 100 million. (Scientists have given a name to about 1 million of them, and of those named, only about 10 percent have been studied in any detail.) Throughout the earth's history, new species have evolved and others have become extinct, with the general trend being that more new species are created than die out. It is now believed that because of human actions this trend has been reversed, with extinctions outnumbering the creation of new species. And the trend appears to be increasing.

According to Edward O. Wilson of Harvard University, probably the most respected of all US biologists, the world has experienced five major periods, or "spasms," of extinction of large numbers of species, from which it took millions of years to recover. These extinctions were caused by natural forces, such as a change of climate. Wilson believes that because of the vast growth of the human population and the related widespread deforestation and overuse of grasslands that is now occurring on our planet, the earth is heading into the sixth and worst period of extinction of species. Wilson estimates the present rate of extinction as about 27,000 species per year, or three per hour. (The normal "background" rate is about 10 to 100 per year). If the present rate continues, Wilson estimates that 20 percent of all the species in the world will be extinct in 30 years.[84]

Whereas hunting used to be the main way humans caused extinction, it is now generally believed that the destruction of natural habitats is the principal cause of extinctions. As the human population grows, humans exploit new areas of the world for economic gain and often destroy life forms as they do so. Biologists believe that about one-half of all species live in tropical forests, which as we have seen above are being cut down at an increasing rate.

A dramatic example of how habitats are destroyed can be seen by looking at a large land development scheme known as "Jari" in the Amazon valley. The Amazon is the largest tropical rain forest on earth. The year-round warm temperature, heavy rainfall, and abundant sunlight, produce excellent conditions for the evolution of species. Species can be destroyed, however, when the land is cleared to make way for farms and commercial enterprises.

In the late 1960s, the US shipping executive and financier Daniel Ludwig, one of the richest persons in the United States, purchased Jari, a parcel of land in the Brazilian Amazon approximately the size of the state of Connecticut. Ludwig invested about $1 billion to construct a paper-pulp factory there. (The factory and a wood-burning power plant were constructed in Japan and towed to the Amazon on huge barges.) Large parts of the forest on Ludwig's estate were cut down and burned to make way for the planting of two or three species of fast-growing trees he brought into the area. As Ludwig said, "I always wanted to plant trees like rows of corn."[85] Ludwig got his rows of trees, but he probably also caused the extinction of an unknown number of plants, insects, and animals.[86] One author described what it was like to walk through one of the new forests at Jari: "no snakes lurked beneath the log, no birds sang in the branches, and no insects buzzed in the still air."[87]

Jari is unique because of the large size of the undertaking, but other smaller developments are becoming more and more common in the remaining rain forests in Latin America, Africa, and Southeast Asia. Scientists fear that the extinctions that these developments are causing could be a direct threat to the well being of human as well as other life on earth.

Many of the species in the tropics have never been studied by scientists. But based on past experience, it is believed that many of these unknown species contain properties that could directly benefit humans. Nearly one-half of the prescription drugs now sold in the United States have a natural component in them.[88] The importance of some of these drugs can be illustrated by the example of just one plant from the tropical rain forests, the rosy periwinkle. Drugs are now produced from this plant that achieve 80 percent remission in leukemia and Hodgkin's disease patients.

Exotic species are vital to the health of modern agriculture. The wild varieties and locally developed strains of a number of major grains grown today have characteristics that are of vital importance to modern seed producers. Seeds are needed with natural resistances to the diseases and pests that constantly threaten modern agriculture. Many farmers today utilize only a relatively few, highly productive , varieties of seeds in any one year. The monocultures that are planted are especially vulnerable to diseases and to pests that have developed resistance to the pesticides being used. An example of how this works was shown in 1970, when 15 percent of the corn crop in the United States was killed by a leaf disease, causing a $2 billion loss to farmers and indirectly to consumers because of higher prices. That year, 70 percent of the corn crop used seeds from only five lines of corn. The disease was finally brought under control with the aid of a new variety of corn that was resistant to the leaf disease. The new corn had genetic materials originating in Mexico.[89]

Insects from tropical forests can at times prove extremely valuable to American farmers. Citrus growers in the United States saved about $25-

$30 million a year with the one-time introduction from the tropics of three parasitic wasps that reproduced and preyed on the pests attacking the citrus fruit.[90] (The introduction of exotic species by humans for profit, or amusement, or by accident into areas to which they are not native is now recognized as having great potential for harm. Since the new species usually has no natural predators in the new area, it can multiply rapidly, destroying or displacing other desirable animals or plants, as was the case with the introduction of rabbits in Australia, and starlings and the kudzu plant in the United States.)

Biologists Paul and Anne Ehrlich outline three main ways the trend toward increased extinction of species can be reversed. First, human population control is urgently needed in many parts of the world, since it is excessive population pressure that is leading to the destruction of habitats in many cases. Second, large reserves should be set up, in carefully selected areas around the world, so species can be preserved in their natural settings. Third, a sustainable society (the characteristics of a sustainable society will be examined fully in chapter 7) should be created. The Ehrlichs define a sustainable society "as one dedicated to living within environmental constraints rather than perpetually growing with the hopeless goal of conquering nature."[91]

Paul and Anne Ehrlich call upon the rich nations of the world to reverse their "overdevelopment," and to help the poorer nations preserve tropical habitats by giving them aid and enacting new international trade policies. The Ehrlichs do not believe that developing nations can preserve tropical habitats on their own since their financial needs are so great. What is needed in the world, they feel, is a new awareness that the diversity of life forms on earth is a priceless treasure that all humanity benefits from and that all share a responsibility for helping to preserve.[92]

The Extinction of Cultures

There are about 15,000 nations on our planet and about 200 nation-states. The nation-states are the political entities, what are commonly referred to as countries. They are often made up of several or many individual nations, or different cultures. The nation is a group of people that share a common history, a common ancestry, and usually a common language and a common religion. They often have common traditions, common ways of doing certain things and of interacting with each other and toward outsiders. Because of these common features that make them different from other peoples, each nation's people see the world and their place in it differently than others, approach problems differently, and have arrived at different solutions to situations humans face. The unique language of the culture is used to pass the common history and traditions down to the young. Linguists now estimate that of the 6,000 languages in the world, one half of them are endangered – that is, like species, they are in danger of extinction.[93] The group that speaks the endangered language – which is often unwritten – is becoming so small that there is a real possibility the group will die out or become absorbed by the larger dominant culture around it and will disappear forever.

Should we care? What will be lost if a culture dies out? The answer to that question is in some ways similar to the answer this book has given to the growing extinction of species. Species represent the amazing variety of life forms on this planet. Their interrelationships are still imperfectly known – to put it mildly – and that can affect the health, and even survival, of one of the species, our own. Cultures also represent the amazing variety of life on earth. But here it is not the form of life that is different, but the different ways members of one species – the human species – have created to live. The culture represents the accumulated knowledge of one group, knowledge that is available to others to pick and choose from, so they can improve their own lives. In addition, as with species, the multitude of cultures makes life on earth extremely rich and varied. The discovery of that variety often leaves an observer with a sense of awe and with a realization that the death of any species or culture leaves life a little less wonderful.

Development, especially since World War II, has often been equated with the culture of the United States. The United States is the largest producer of goods and services and its culture is closely associated with material wealth. Freedom from the burdens of excessive control by government and freedom from the restrictions common in more traditional societies are also characteristics of US society. These characteristics have contributed to an emphasis on innovation and change that has led to many new products and services. So it is not surprising that development and US culture have seemed to go together. Many other cultures have found that their youth are more attracted to the US culture than to their own. So also within the United States, ethnic groups have found that it is extremely difficult to keep from being absorbed by the dominant culture. The youth of the ethnic groups want to become accepted by the majority and they know that this will come only if they are like the majority, not different from it.

Although the United States has often been called a "melting pot," a country where immigrants from many different countries and with many different cultures melted or blended into one culture, some people have gradually come to see that what is really going on is a destruction of the rich diversity in the country and the creation of a rather boring homogeneity. There is a growing recognition that a high price is being paid for the understandable need for unity in the country, for the feeling that all these people from different backgrounds are now American.

For some people it is comforting to find a McDonald's fast food restaurant or a Wal-Mart department store no matter where you go in the country, but to others this spreading sameness has meant a loss of the rich regional differences that have made the country fascinating. The interstate highway system symbolizes well the strengths but also the weaknesses of that homogeneity. It is a very efficient way to travel by car, but if one wants to experience the variety existing in the country, one has to leave it and drive through the small towns located on the smaller roads.

Some people in the United States, as well as in other developed countries are giving a new respect to what had formerly been labeled as "primitive" cultures. There is a slowly growing recognition that these cultures may have knowledge that developed countries need if they are going to survive – such as an ability to live in harmony with nature, a concern for future generations, and a knowledge of how to foster a sense of community.[94] Tribal people in tropical forests have been finally recognized as possessors of important knowledge regarding natural drugs in plants and of skills that have enabled them to live in the forests without destroying them. There is also a growing recognition that if you want to preserve the tropical forests and the multitude of species that they harbor, you must make it possible for the indigenous people living in them to survive. If these people cannot survive, probably the forests cannot either. If these people do survive, they will help protect the forests that are their homes.

Let us now focus for a moment on two cultures under stress at present and in danger of extinction. One culture, the Yanomami, is found in two developing countries – Brazil and Venezuela – and the other, the Estonians, is found in Europe.

The Yanomami

In the Amazon region of Latin America live the Yanomami. It is believed that these people have lived in this region for thousands of years.[95] The approximately 20,000 Yanomami represent the largest group of indigenous people living in the Americas who still follow Stone Age methods. Although they had very limited contact with other peoples for many years, this changed in the late 1980s when gold was discovered in the Brazilian Amazon region. Thousands of miners flew into the area where the Yanomami lived. The miners brought with them diseases to which the Yanomami had no natural immunity. Amnesty International estimates that from 1988 to 1990 about 1,500 Yanomami died.[96] In addition to the malaria that killed many, some Yanomami died from mercury poisoning, which came from eating fish poisoned by the mercury the miners had used in the streams to sift for gold. Others were killed by armed attack. Amnesty International reported that "These attacks are often carried out by private agents, including gunmen hired by land claimants, timber merchants or mining interests. They have gone almost entirely unpunished – in fact, state-level authorities have even colluded with them."[97]

The Yanomami's situation became known throughout Brazil and around the world. Responding to pressures within Brazil and from some foreign countries (the attention given to Brazil because of the upcoming United Nations environmental conference probably played a role), the Brazilian government in 1991 set aside for the Yanomami about 36,000 square miles of land. When added to that set aside by Venezuela, which was slightly smaller than the Brazilian grant, this was an amount of land equal to the size of Portugal and the amount anthropologists said the Yanomami needed in order to survive. In 1993 Brazil used its police and military force to forcibly remove 3,000 miners who were still in Yanomami lands.

What is the fate of the Yanomami? No one knows, of course, but if history is a guide, one would have to say that their prospects of surviving are not bright. While the actions by the Brazilian and Venezuelan governments to reserve a large amount of land for the use of these people is a hopeful step, disturbing signs exist. The presence of gold in their lands is unfortunate. In 1990 the agency in charge of Indian affairs in the Brazilian government announced that it was forcibly removing *all* miners from Yanomami lands.[98] But in 1993, as mentioned above, 3,000 miners were still there. Any attempt to keep the miners out permanently will probably fail. The other disturbing fact is that there is abundant research now showing that when indigenous peoples come in contact with the modern world, they often lose the special knowledge possessed by members of their culture – such as natural healing methods and drugs – and develop a dependency on modern goods, which destroys their self-sufficiency and pride.[99] Suicide rates and alcoholism often soar.

The Estonians

The Estonians live in a tiny country in northern Europe. With less than 1 million people, there are fewer Estonians than the population of most major cities in the world.

Their culture is a very old one. Estonians have lived in this same spot next to the Baltic Sea for 5,000 years. Many foreigners have ruled the land during the past 700 years – among them Swedes, Germans, and Russians – but Estonian culture has survived. Now, after having survived the latest foreign rule – an especially hard 50 years under the Soviet Union – Estonians are struggling to keep their culture alive. The country has a slightly negative rate of population growth, a situation that could be dangerous if it continues too long.

Estonians have survived in spite of efforts by their latest rulers to destroy their culture. As they did to other peoples they conquered, the Soviets shipped many Estonian intellectual, political, and military leaders to Siberia. Others were killed within the country. Stalin introduced measures to Russify the country. One way he tried to do this was to introduce heavy industry into Estonia and to import Russian workers to run the plants. In the latter part of the twentieth century, Estonians were aware that if they did not regain their independence soon, their culture would be destroyed.

The way Estonians regained their independence marks them as a very unusual people. When I visited the country in 1990 they were still under Soviet rule. As a political scientist I knew that their chances of winning back the independence they had enjoyed before Hitler and Stalin made a deal in 1939 to give the country to the Soviet Union were very slim. US political scientists knew that the Soviets could not agree to Estonian independence without threatening the very foundations of the Soviet Union itself. Yet within one year after my visit, Estonia had become independent. It did this through nonviolent opposition and by waiting for the right moment to declare its freedom. When a coup d'état was attempted in the Soviet Union by conservative forces in 1991, the Estonians moved. That move was followed by similar declarations of independence by the two other small Baltic nations of Lithuania and Latvia. Not long after that the Soviet Union itself broke up.

The Estonian fight to regain their independence has been called the "Singing Revolution."[100] Instead of using guns to push out the Soviets – an effort the Soviets probably would have welcomed so they would have had an excuse to crush the independence movement – the Estonians used songs. They had persuaded the Soviets to build a huge outdoor stadium in Tallinn, the Estonian capital city, where their song festivals – an important part of the Estonian culture – could be held. What the Russians didn't realize was that the song festivals were helping to keep alive the Estonians' love for their land and freedom.

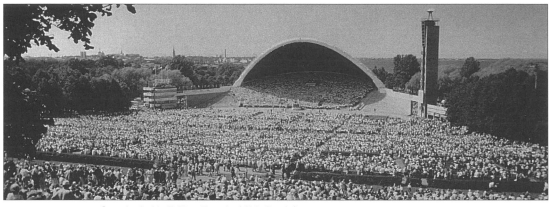

Song festival in Tallinn, Estonia. (*Aldo Bender*)

The 1990 song festival in Estonia was the last one under Soviet rule. I was fortunate enough to attend it and found it to be an extremely moving experience. The two-day festival, the largest song festival in the world, was attended by about 500,000 people – about one-half of the nation. The 28,000 singers from all parts of the country, dressed in traditional clothes, sang of their love for their homeland and their desire to be free. Nine thousand dancers performed in colorful costumes, each unique to a different section of the country.

Two years after regaining its independence Estonia was being cited in the Western press as a model for the rest of the former republics of the Soviet Union to follow as they tried to pass from stagnant, centrally planned economies to those based on individual initiative and freedom. Today the country has made relatively good economic progress, but huge problems still exist. After many years of Soviet control that were designed to replace love for and loyalty to their own culture with loyalty to the Soviet state (with its supposedly new type of person), the Estonians have an uphill battle. Many Russians, who have never learned the Estonian language or participated in the culture, still live in Estonia. And in the early 1990s the Russian army was still there. (It finally left in 1994.)

Even if the threat from Russia recedes, the Estonians face another threat to their culture. As they develop and become part of the Western world, they fear that the dominant US-influenced culture will come to replace their own. They face the same challenge as do other nations with their distinctive cultures. Can they develop but yet retain their distinctiveness, their own culture? Or will the influences brought by the new opportunities to travel, increased contact with Western goods and tourists, and messages they receive from the Western media overwhelm their own ways? Cultures are always changing and this can be healthy. But can Estonia change in some ways but not change in others? Will they continue to be Estonians? Only time will tell.

Environmental Politics ●

In this final section we shall try to understand what makes environmental politics so controversial. Politics is a passionate business, but why are environmental issues often emotional? Obviously, conflicting interests and values must be involved. Politics involves the making of laws and decisions that everyone must obey in a society. These laws and decisions are directed at settling conflicts that arise among people living together in a community, and at achieving commonly desired goals. As we shall see, environmental politics does deal with very strongly held opposing values and interests. It also represents an effort by a community to achieve some goals – such as clean air and clean water – which cannot be reached individually, only by the community as a whole.

The political scientists Harold and Margaret Sprout believe that most participants in environmental politics show a tendency toward having one of two very different philosophies or world views and that these are at the root of most environmental conflicts. One they call "exploitive," and the other "mutualistic." Here is how they define them:

A(n)...exploitive attitude would be one that envisages inert matter, nonhuman species, and even humans as objects to be possessed or manipulated to suit the purposes of the exploiter. In contrast, a...mutualistic posture would be one that emphasizes the interrelatedness of things and manifests a preference for cooperation and accommodation rather than conflict and domination.[101]

While conflicting world views are a part of environmental politics, so also is a conflict of basic interests. Economist Lester Thurow believes that environmental politics often involves a conflict between different classes having very different interests. He sees the environmental movement being supported mainly by upper-middle-class people who have gained economic security and now want to improve the quality of their lives further by reducing environmental pollutants. On the opposite side, he sees both lower-income groups and the rich – lower-income people because they see environmental laws making it more difficult for them to find jobs and obtain a better income, and the rich because they can often buy their way out of environmental problems and see pollution laws as making it more difficult for them to increase their wealth even further.[102]

Other conflicting interests are also involved in environmental politics. Antipollution laws often make it more difficult and costly to increase energy supplies, extract minerals, and increase jobs by industrial growth. Barry Commoner's Fourth Law of Ecology – There Is No Such Thing as a Free Lunch – means that for every gain there is some cost.[103] There are tradeoffs involved in making the air and water cleaner as there are in making more cars and television sets. Also, the costs of pollution control often increase substantially as you try to make the environment cleaner and cleaner. The cost required to make a 50 percent reduction in a pollutant is often quite modest, whereas if you try to reduce the pollutant by 95 percent, the cost usually increases dramatically.[104]

Much environmental destruction is extremely difficult for the political system to deal with, since the damage often shows up many years after the polluting action takes place. It is now clear that prevention is much cheaper than trying to clean up the damage after it has occurred, but the nature of politics does not lend itself to long-range planning. Generally, politicians have a rather short-term outlook, as do many business people. Both are judged on their performance in handling immediate problems; this promotes a tendency to take actions showing some immediate result. Such actions further the politician's chances for reelection and the business person's profits or chances for promotion. Yet environmental problems often call for actions before the danger becomes clear. A further complication is the fact that, even after action is taken to reduce a pollutant, because of the inherent delays in the system, the harmful effects of the pollutant do not decrease until a number of years later. Thus the inclination of the public official – and the business person – is to do nothing and hope that something turns up showing that the problem was not as bad as feared or that there is a cheaper way to deal with it.

An additional factor in environmental politics is unique to the United States. The American dream has been one of continuing abundance. For much of the country's history, there has seemed to be an unlimited abundance of many things needed for the good life, such as land, forests, minerals, energy, clean air, and natural beauty. It is a country that seemed to offer unlimited opportunities for many to a make a better life for themselves, and "better" has been usually defined as including more material goods. The setting of limits on consumption and production that environmentalists often promote is certain to cause dismay to many.

If the above were not enough to make environmental politics very difficult, there is also the fact that the costs in environmental matters are often very difficult to measure. One can calculate the cost of a scrubber on a coal-burning power plant, but how do you measure the cost of a shortened life that occurs if the scrubber is not used? How do you place a dollar figure on the suffering a person with emphysema experiences, or a miner with brown lung disease, or an asbestos worker with cancer? How do you measure the costs the yet unborn will have to pay if nothing is done now about acid rain? And how do you put a dollar figure on the loss of natural beauty? Because it is so difficult to weigh the costs in conventional terms of measurement, the costs often were not weighed in the past.[105]

There is, of course, also the matter of values – the value individuals place on more material goods, the convenience of throwaway products, open spaces, and clean air. The resolution of conflicts over values can often be handled only by politics, in a democracy by the community as a whole making decisions through its representatives and then requiring all members of the community to obey them. That such stuff causes controversy and stirs passions should not be surprising. It is hard work.

Conclusions

Development is more than economic growth: it also includes the social changes that are caused by or accompany economic growth. As this chapter has shown, the increase in the production of goods and services that came with industrialization had, and still has, frightening costs. Poverty was basically wiped out in a number of countries by industrialization – obviously an impressive benefit of the new economic activity. But that activity harmed both people and the environment. Slowly and painfully, people in the developed countries have come to realize that economic growth is not enough. Attention has to be paid to its effect on the earth and on people. (If one gets cancer, for example, what good is material wealth?) And an awareness has grown in the industrialized nations, and continues to grow, that the question of how economic growth is affecting the environment needs to be asked and answered. The rich countries are slowly learning that it is cheaper and causes much less suffering to try to reduce the harmful effects of an economic activity at the beginning, when it is planned, than after the damage appears. To do this is not

175

easy and is always imperfect. But an awareness of the need for such effort indicates a greater understanding and moral concern than did the previous widespread attitude that focused only on creating new products and services.

The less developed nations are also slowly realizing that the effects of economic activity on the environment should not be ignored. But here the new awareness is less widespread than in the rich countries. This is understandable because, except for some of the rulers and elite groups, the reduction of poverty is the first concern people have. It explains why some Third World countries have welcomed polluting industries, such as factories that manufacture asbestos, since jobs today are more important than a vague worry that workers may contract cancer in 20 to 30 years. But also in developing countries, a slowly growing number of people realize that if the economic activity that gives jobs to people harms the environment at the same time, the benefits from that economic activity will be short-lived.

Poverty harms the environment, as we saw for example in the case of deforestation, where poor people searching for land to farm and for fuel are one cause of the extensive destruction of the remaining tropical rain forests. Economic growth that benefits the majority of people is needed to protect the environment. And a control of the rapidly expanding populations of many of the poorest countries is also needed to protect the environment, since increasing numbers of poor people hurt the land on which they live as they struggle to survive.

For both rich and poor nations, the environment is important. Economic growth is also important, especially for the poorer countries. The challenge remains for both poor and rich to achieve a balance between economic activity and a protection of the land, air, and water upon which life depends.

Notes

1   Bill Clinton, who defeated President Bush, stated he would sign the biodiversity treaty.
2   Erik P. Eckholm, *Down to Earth: Environment and Human Needs* (New York: W. W. Norton, 1982), p. 95.
3   Lester Brown, *The Twenty-ninth Day* (New York: W. W. Norton, 1978), p. 44. One of the ways Britain reduced its air pollution was to build tall smokestacks, which has probably led to worse air in Scandinavia.
4   Stephen Klaidman, "Muddling Through," *Wilson Quarterly* (Spring 1991), p. 76.
5   Jon Luoma, "Sharp Decline Found in Arctic Air Pollution," *New York Times*, national edn (June 1, 1993), p. B7.
6   "Air Found Cleaner in 41 U.S. Cities," *New York Times*, national edn (October 20, 1992), p. A17.
7   "Big Cities Still Lagging on Clean Air, U.S. Says," *New York Times*, national edn (August 17, 1990), p. A10.
8   Philip Hilts, "Studies Say Soot Kills Up To 60,000 in U.S. Each Year," *New York Times*, national edn (July 19, 1993), p. A1.
9   World Bank, *World Development Report 1992: Development and the*

Environment (New York: Oxford University Press, 1992), pp. 50–2.

10 Marlise Simons, "Rising Iron Curtain Exposes Haunting Veil of Polluted Air," *New York Times*, national edn (April 8, 1990), p. 1.

11 *New York Times*, late city edn (May 13, 1980), p. C3.

12 Jane Brody, "Lead-Poisoning Harm Held to Be Partly Reversible," *New York Times*, national edn (April 8, 1993). A 1993 study revealed that the decline in intelligence caused by lead poisoning could be partly reversed by lowering the lead in the blood.

13 Sandra Blakeslee, "Study Shows a Steep Drop of Levels of Lead in Blood," *New York Times*, national edn (July 27, 1994), p. C20. For some unknown reason black children seem to be more susceptible to lead poisoning than white children. See Jane Brody, "Despite Reductions in Exposure, Lead Remains a Danger to Children," *New York Times*, national edn (March 21, 1995), p. B7.

14 *New York Times*, national edn (July 27, 1994), p. C20.

15 "Lead Concentrations Down in Greenland Ice," *New York Times*, national edn (October 15, 1991), P. B8.

16 World Bank, *World Development Report 1992*, p. 53.

17 Marlise Simons, "Pollution's Toll in Eastern Europe: Stumps Where Great Trees Once Grew," *New York Times*, national edn (March 19, 1990), p. A9. An in-depth analysis of the environmental crisis in Czechoslovakia is contained in Sam Bingham, "Czechoslovakian Landscapes," *Audubon* (January 1991), pp. 92–103.

18 Barry Commoner, *The Closing Circle* (New York: Alfred A. Knopf, 1971), p. 39.

19 *New York Times*, late edn (June 30, 1983), p. A16.

20 William Stevens, "Worst Fears on Acid Rain Unrealized," *New York Times*, national edn (February 20, 1990), p. B5; and Philip Shabecoff, "Acid Rain Report Confirms Concern, but Crisis Is Discounted," *New York Times*, national edn (September 6, 1990), p. A14.

21 William Stevens, "What's a Little Pollution? Europe's Forests Keep On Growing," *New York Times*, national edn (April 7, 1992), p. B7.

22 William Stevens, "Ozone Loss Over U.S. Is Found To Be Twice as Bad as Predicted," *New York Times*, national edn (April 5, 1991), p. A1.

23 William Stevens, "Summertime Harm To Shield of Ozone Detected Over U.S.," *New York Times*, national edn (October 23, 1991), p. A1.

24 Owen B. Toon and Richard P. Turco, "Polar Stratospheric Clouds and Ozone Depletion," *Scientific American*, 264 (June 1991), p. 74.

25 William Stevens, "Ozone Layer Thinner, But Forces Are in Place for Slow Improvement," *New York Times*, national edn (April 9, 1991), p. B7.

26 *New York Times*, late edn (March 28, 1983), p. 1.

27 World Resources Institute, *The 1993 Information Please Environmental Almanac* (Boston: Houghton Mifflin, 1993), pp. 38–40.

28 Brown, *The Twenty-ninth Day*, p. 40.

29 Hilary F. French, "Restoring the East European and Soviet Environments." In Lester R. Brown et al., *State of the World–1991* (New York: W. W. Norton, 1991), p. 97.

30 Lester R. Brown, "A New Era Unfolds" In Lester R. Brown et al., *State of the World–1993* (New York: W. W. Norton, 1993), p. 10.

31 US Environmental Protection Agency, *Progress in Groundwater Protection and Restoration* (Washington: US Environmental Protection Agency, 1990), p. 18.

32 World Resources Institute, *World Resources 1992–93* (Oxford: Oxford University Press, 1992), p. 103.

33 Ibid.

34 Stephen Klaidman, "Muddling Through," p. 76.

35 Peter Stoler, "Is Clean Water a Thing of the Past?" *Sierra* (March /April 1981), p. 14.

36 Joyce Egginton, "The Long Island Lesson," *Audubon*, 83 (July 1981), p. 89.

37 Council on Environmental Quality and the Department of State, *The Global 2000 Report to the President: Entering the Twenty-First Century*, vol. 1 (New York: Penguin Books, 1982), p. 26.

38 Stephen Klaidman, "Muddling Through," pp. 76–7.

39 Peter Passell, "The Garbage Problem: It May Be Politics, Not Nature," *New York Times*, national edn (February 26, 1991), p. B7.

40 See Adeline Levine, *Love Canal: Science, Politics and People* (Lexington, Mass.: Lexington Books, 1982) and Michael Brown, *Laying Waste: The Poisoning of America by Toxic Wastes* (New York: Pantheon Books, 1980).

41 Scott Pendleton, "Incineration Expected to Increase in '90s," *Christian Science Monitor* (April 25, 1991), p. 12.

42 Keith Schneider, "Release of Toxic Waste Declines As Industry Recycles More of It," *New York Times*, national edn (May 26, 1993).

43 *New York Times*, late edn (March 11, 1985), p. D12.

44 "'Green' Goes Grass Roots," *Christian Science Monitor*, (January 12, 1993), p. 10.

45 Barbara Ward, *Progress for a Small Planet* (New York: W. W. Norton, 1979), pp. 65–6.

46 Mary Durant, "Here We Go A-Bottling," *Audubon*, 88 (May 1986), p. 32.

47 Michael Royston, "Making Pollution Prevention Pay," *Harvard Business Review* (November-December 1980), p. 12.

48 William Booth, "Tropical Forests Disappearing at Faster Rate," *Washington Post* (September 9, 1991), p. A18.

49 *New York Times*, national edn (May 5, 1992), p. B7.

50 Ibid.

51 *New York Times*, late edn (July 5, 1983), p. C1.

52 Norman Myers, *The Primary Source* (New York: W. W. Norton, 1984), p. 143. Local government officials can be directly involved in the logging industry for private gain. In Thailand, the Philippines, and other Southeast Asian countries, cabinet ministers, senators, and other senior politicians are involved in the timber industry. In Indonesia most of the holders of extensive timber concessions are retired military and government officials. See Robert Repetto, "Deforestation in the Tropics," *Scientific American*, 262 (April 1990), p. 39.

53 Anne LaBastille, "Heaven, Not Hell," *Audubon*, 81 (November 1979), p. 91.

54 *New York Times*, late edn (January 24, 1984), p. C2.

55 Christopher Uhl, "You Can Keep a Good Forest Down," *Natural History*, 92 (April 1983), p. 78.

56 William Stevens, "Shamans and Scientists Seek Cures in Plants," *New York Times*, national edn (January 28, 1992), p. B5.

57 Ghillean Prance, "Fruits of the Rain Forest," *New Scientist* (January 13, 1990), pp. 42-5; and Catherine Dold, "Tropical Forests Found More Valuable for Medicine Than Other Uses," *New York Times*, national edn (April 28, 1992), p. B8. Some research has also shown that efforts to market products from the forests can bring harmful changes to the local people. See Andrew Gray, "Indigenous Peoples and the Marketing of the Rain Forest," *The Ecologist*, 20 (November/December 1990), pp. 223–7; and Katharine Milton, "Civilization and Its Discontents," *Natural History* (March 1992), pp. 37–43.

58 Brown, *The Twenty-ninth Day*, p. 57.

59 Richard Peto, "Why Cancer?" In Julian Simon and Herman Kahn (eds) *The Resourceful Earth* (Oxford, England: Basil Blackwell, 1984), pp. 528–44; and *New York Times*, late edn (March 20, 1984), p. C1.

60 Ted Williams, "Hard News on 'Soft' Pesticides," *Audubon*, 95 (March-April 1993), pp. 30–40.

61 "Study Links DDT and Cancer," *New York Times*, national edn (April 22, 1993), p. A10.

62 Ibid.

63 David Weir and Mark Schapiro, *Circle of Poison* (San Francisco: Institute for Food and Development Policy, 1981), p. 4.

64 Ibid.

65 *New York Times*, late edn (April 21, 1986), p. A14.

66 Philip Hilts, "Results of Study on Pesticide Encourage Effort to Cut Use," *New York Times*, national edn (July 5, 1993), p. 8.

67 *New York Times*, late edn (March 3, 1984), p. 10.

68 Commoner, *The Closing Circle*, p. 261.

69 Hans Landsberg et al., "Nonfuel Minerals," In Paul Portney (ed.), *Current Issues in Natural Resource Policy*, (Washington: Resources for the Future, 1982), p. 82.

70 Ibid., p. 83.

71 Dennis Pirages, *The New Context for International Relations: Global Ecopolitics* (North Scituate, Mass.: Duxbury Press, 1978), p. 172.

72 Marlise Simons, "Dutch Feel Burden of U.S. Recycling," *New York*

Times, national edn (December 11, 1990), p. A6.

73 Jon R. Luoma, "Trash Can Realities," *Audubon* (March 1990), p. 93.

74 Laurent Belsie, "Recycling Abounds Across America," *Christian Science Monitor* (July 18, 1990), p. 8.

75 *Christian Science Monitor* (April 30, 1992), p. 7.

76 As quoted in Brown et al., *State of the World–1993*, p. 192.

77 Jeremy Rifkin, *Entropy* (New York: Viking Press, 1980), p. 117.

78 Luoma, "Trash Can Realities," p. 95.

79 Brown, *The Twenty-ninth Day*, p. 284.

80 Elizabeth Brown, "Bottle Bills Proliferate in States and in Congress," *Christian Science Monitor* (March 5, 1991), p. 7.

81 Ted Williams, "The Metamorphosis of Keep America Beautiful," *Audubon* (March 1990), p. 132.

82 Ibid., pp. 128–9.

83 Charles Birch, *Confronting the Future* (New York: Penguin, 1976), p. 35.

84 Edward O. Wilson, *The Diversity of Life* (Cambridge: Harvard University Press, 1992).

85 Loren McIntyre, "Jari: A Billion Dollar Gamble," *National Geographic* (May 1980), p. 701.

86 After 14 years of building Jari, Ludwig abandoned the project in 1982 and sold it to Brazilians for a loss. Ten years after the sale, some were claiming that the new Brazilian owners of Jari had turned it into a successful example of sustainable development in the Amazon. A fast-growing Eucalyptus tree was being harvested for wood pulp. Virgin forests were left around the sections used for tree farming in order to prevent erosion and allow enough biodiversity to help keep pests under control. It is still too early to make a final judgment of this scheme. See Jeb Blount, "Brazil Tree Farm Uses Rain Forest and Also Saves It," *Christian Science Monitor* (April 27, 1993), pp. 10–11.

87 McIntyre, "Jari," p. 711.

88 Peter Raven, "Tropical Rain Forests: A Global Responsibility," *Natural History*, 90 (February 1981), p. 29.

89 Norman Myers, "The Exhausted Earth," *Foreign Policy*, 42 (Spring 1981), p. 143.

90 Norman Myers, "Room in the Ark?" *Bulletin of the Atomic Scientists*, 38 (November 1982), p. 48.

91 Paul R. Ehrlich and Anne H. Ehrlich, *Extinction* (New York: Random House, 1981), pp. 242–3.

92 Ibid., especially ch. 10. Note their following statement: "Over 95 percent of the organisms capable of competing seriously with humanity for food or of doing us harm by transmitting diseases are now controlled gratis by other species in natural ecosystems" (p. 94).

93 Philip E. Ross, "Hard Words: Trends in Linguistics" *Scientific American*, 264 (April 1991), p. 145.

94 As one author has written: "[indigenous peoples] may offer living examples of cultural patterns that can help revive ancient values for everyone: devotion to future generations, ethical regard for nature, and commitment to community among people." Alan Durning, "Supporting Indigenous Peoples." In Lester Brown et al., *State of the World–1993* (New York: W. W. Norton, 1993), p. 100.

95 Pottery has been found in the lower Amazon Basin which is 7,000 to 8,000 years old, the oldest pottery found in the Americas. John Wilford, "Oldest Pottery in Americas Is Found in Amazon Basin," *New York Times*, national edn (December 13, 1991), p. A13.

96 James Brooke, "Brazil Evicting Miners in Amazon to Reclaim Land for the Indians," *New York Times*, national edn (March 8, 1993), p. A4.

97 Ibid.

98 James Brooke, "In an Almost Untouched Jungle Gold Miners Threaten Indian Ways," *New York Times*, national edn (September 18, 1990), p. B6.

99 One encouraging sign is that when Western scientists seek information from the medicine men of indigenous peoples about natural drugs and health cures, the medicine men are given new respect. This new respect might help encourage some of their youth to study under them. But, all too often today, when the medicine men die, the knowledge they have acquired dies with them. See Daniel Goleman, "Shamans and Their Longtime Lore May Vanish With the Forests," *New York Times*, national edn (June 11, 1991), p. B5. As an example of a study showing the harmful effects Western contact can have on the culture of indigenous peoples, see Katharine Milton, "Civilization and Its Discontents."

100 See, for example, Walter C. Clemens Jr., "Baltics Sang Their Way to Independence," *Christian Science Monitor* (September 5, 1991), p. 19.

101 Harold Sprout and Margaret Sprout, *The Context of Environmental Politics* (Lexington, Ky.: University Press of Kentucky, 1978), pp. 47–8.

102 Lester Thurow, *The Zero-sum Society* (New York: Basic Books, 1980), pp. 104–5.

103 Commoner, *The Closing Circle*, pp. 45–6.

104 William Ophuls, *Ecology and the Politics of Scarcity* (San Francisco: W. H. Freeman, 1977), p. 75.

105 For an attempt to measure the hidden costs of today's energy, see Harold M. Hubbard, "The Real Cost of Energy," *Scientific American*, 264 (April 1991), pp. 36–42. For an explanation of how the accounting system presently used by economists allows policy makers to ignore the deterioration of the environment caused by economic activity, see Robert Repetto, "Accounting for Environmental Assets," *Scientific American*, 266 (June 1992), pp. 94–100.

Further Readings

Adams, Jonathan S., and Thomas O. McShane, *The Myth of Wild Africa: Conservation Without Illusion* (New York: W.W. Norton, 1992). The central argument in this book is that it is impossible to protect the animals of Africa unless the people living around them are given an economic incentive to help preserve the animals.

Benedick, Richard Eliot, *Ozone Diplomacy: New Directions In Safeguarding the Planet* (Cambridge: Harvard University Press, 1991). The head of the US delegation negotiating the Montreal Protocol on Substances that Deplete the Ozone Layer argues that the government must take action on environmental issues despite scientific uncertainty rather than risk the consequences of inactivity.

Brown, Phil, and Edwin J. Mikkelsen, *No Safe Place: Toxic Waste, Leukemia, and Community Action* (Berkeley: University of California Press, 1990). This book documents the efforts of a small group of citizens – led by a mother whose son had acute leukemia – to uncover the source of illness in their community. It describes their battle to overcome resistance by their neighbors and the indifference of local and state government officials.

Commoner, Barry, *Making Peace with the Planet* (New York: Pantheon Books, 1990). The author calls for new and generally small-scale technologies (based on solar energy and massive recycling) to help people live in a way that will not harm the planet. While documenting the failure of governments and corporations to control environmental dangers, he argues that prevention of pollution is needed rather than an attempt to control it.

Coswell, Adrian, *The Decade of Destruction* (New York: Doubleday, 1991). A personal account of the destruction of the Amazon rain forests illustrates the complexities of the issue and the value of the forests.

Feshbach, Murray and Alfred Friendly, Jr, *Ecocide in the USSR: Health and Nature Under Siege* (New York : Basic Books, 1992). In the process of becoming an industrial power, the Soviet Union inflicted great harm on the environment and its own people.

Firor, John, *The Changing Atmosphere* (New Haven: Yale University Press, 1990). This clear and readable text discusses the overall impact of human activities on the atmosphere and focuses on specific problems, such as acid rain, ozone depletion, and global warming.

Maybury-Lewis, David, *Millennium: Tribal Wisdom and the Modern World* (New York: Viking Press, 1992). This visually stunning book examines various traditional societies and their store of knowledge, which could be used to benefit the modern world.

McIntyre, Loren, *Amazonia* (San Francisco: Sierra Club Books, 1991). McIntyre has traveled every major tributary in the Amazon Basin. In this book he presents impressive portraits of many of the tribes living there.

McPhee, John, *The Control of Nature* (New York: Farrar, Straus and Giroux, 1989). McPhee documents the puny attempts of humans to control nature, but also some of their successes.

Moyers, Bill D., and the Center for Investigative Reporting, *Global Dumping Ground: the International Traffic in Hazardous Waste* (Washington: Seven Locks Press, 1990). Journalist Moyers investigates toxic dumping, both legal and illegal, the profits and the damage, from Love Canal to the local dump.

Nash, Roderick Frazier, *The Rights of Nature: A History of Environmental Ethics* (Madison: University of Wisconsin Press, 1989). Nash believes that humans, after establishing rights for themselves, gradually expanded the idea of rights to include the family, the tribe, the nation, and in theory, if not in practice, all of humanity. Nash argues that we are now in the process of expanding the concept of ethical and legal rights to cover animals, plants, and the rest of the natural world.

Rathje, William, and Cullen Murphy, *Rubbish! The Archaeology of Garbage* (New York: HarperCollins, 1992). This book is much more than a history of the garbage project (popularly called "garbology") at the University of Arizona, which examined landfills in five US cities, as well as in Mexico and Australia. It also deals with the myths, technologies, and policies of the disposal of solid wastes in modern societies.

Shabecoff, Philip, *A Fierce Green Fire: The American Environmental Movement* (New York: Hill and Wang, 1993). In this history of environmentalism in the United States, the author describes the three waves of environmentalism that have swept over the country.

Wilson, Edward O., *The Diversity of Life* (Cambridge: Harvard University Press, 1992). Wilson shows how the evolution of life has progressed on earth, with five great "spasms" of death occurring along the way. Wilson believes that, for the first time, humans are causing an extinction spasm – the sixth – but he leaves the reader with hope that humans may come to realize what they are doing before it is too late.

6

Technology

Will mankind murder Mother Earth or will he redeem her? He could murder her by misusing his increasing technological potency.

Arnold J. Toynbee (1889 – 1975)
Mankind and Mother Earth (1976)

To many people, technology and development are synonymous. Technology is what makes economic growth and social change happen. The limited use of high technology by the less developed nations is sometimes given as one of the reasons why they are less developed and less prosperous than the industrialized nations. But the relationship between technology and development is a complicated one. At times the negative features of technology seem to outweigh the positive features. Technology can cause a society to change in some very undesirable ways. In this chapter, after a short section on the benefits of technology, we will look closely at some of the negative relationships between technology and development.

Benefits of Technology

A book such as this one, whose readers will probably be mostly from the developed nations, does not need to dwell on the benefits of technology.

Advertising and the mass media herald the expected joys that will come with a new product, technique, or discovery. In the United States people are socialized to like new things; they are also pragmatic, which means that science and its application, technology, are commonly used to solve problems, to make things work "better." They would have to be foolish not to recognize the benefits that technology has brought.

Without modern technology to help, necessary tasks can be difficult. A woman in Nepal breaks up clumps of soil to prepare the land for planting. (*Ab Abercrombie*)

Benefits of Technology

In personal terms, technology has allowed me to visit about 40 countries; to see a photograph of the earth taken from space; to write this book on a personal computer that greatly facilitated its composition; to wear shirts that don't need ironing; and to keep my glaucoma under control so that I do not go blind. What items would your list include?

One of the main reasons much of the world envies the United States is that its technology has in many real ways made life more comfortable, stimulating, and free of drudgery. People in the United States know this and need to remember it. But they and others also need to learn several other lessons: (1) short-term benefits from using a technology can make it impossible to achieve some long-term goals; (2) there can be unanticipated consequences of using a technology; (3) the use of some types of technology in certain situations can be inappropriate; and (4) there are many problems that technology cannot solve. The inability to learn these lessons could lead to our destruction, as the case study in this chapter on the threat of nuclear weapons will show.

Short-term Versus Long-term Benefits ● ● ● ● ● ● ● ● ●

Garrett Hardin, a biologist, has coined the phrase "the tragedy of the commons" to describe what can happen when short-term and long-term interests of people are in conflict.[1] Hardin shows how it is rational and in the best short-term interest of each herdsman in a village to increase the number of cattle he has grazing on the "commons," the open-access commonly owned lands in the village. The apparent short-term benefit to an individual herdsman of increasing the number of cattle he has there is greater than the long-term harm resulting from the overgrazing that the additional cattle create; the cost of the overgrazing will be shared by all the herdsmen using the commons, while the individual herdsman will reap the profit that comes from selling additional cattle. Also, if the individual herdsman does not increase his cattle but others increase theirs, he loses out since the overgrazing harms his cattle. Thus the tragedy occurs. Each herdsman, acting rationally and in his own best short-term interest, increases his stock on the commons. Soon there is so much overgrazing that the grass dies and then the cattle die.

The global commons today are those parts of the planet that are used by many or all nations: the oceans, international river systems, the seabed, the atmosphere, and outer space. Technology can give some nations an advantage over others in exploiting these commons and it is clearly in their short-term interest to do so. So it is with commercial fishing in the world's oceans. Technology has made possible bigger and more powerful fishing boats, equipped with sonar to locate schools of fish. It has also led to the creation of huge drift nets – some up to 40 miles long – which critics claim are being used to "strip mine" the seas. These nets allow a relatively small number of fishermen to catch large quantities of fish. (Efforts are being made in the United Nations to ban the use of these nets.) There is every indication that many species of fish are being "overgrazed," and if this is not controlled, all nations using the oceans for fishing will be hurt. Not only will their fishing industries be hurt, but unique forms of life on earth will probably become extinct. Such could well be the fate of many species of the fish-like mammal, the whale, unless recent international efforts to reduce drastically the numbers of whales killed succeed in allowing whale populations to increase.

An example of technology giving a nation advantages over others in exploiting the global commons can be seen in the history of the Law of the Sea Treaty, which governs the use and exploitation of the seas. This treaty, which took eight years to negotiate, was approved in 1982 by 130 nations at the United Nations. The United States was one of four nations to vote against it. One of the reasons for its opposition was that it was unwilling to share its advanced sea mining technology with a global mining authority. The short-term advantages to the United States of not sharing its mining technology are clear, but these advantages conflict with long-term interests of the United States and other nations in having a peaceful and mutually agreed upon arrangement for the use of this part of the global commons. Donald Puchala summarizes the matter well:

> The United States would probably benefit from a short-run scramble to close off the commons and parcel it into national jurisdictions. Since our technology permits us to exploit now what others can only hope to exploit in the future, we would for a time command the lion's share of the parceled commons. But there should be no doubt that such a policy would invite challenge and conflict in the future....[2]

Another example of a tragedy-of-the-commons situation is pollution. Individuals gain a short-term advantage by polluting – for example, by disconnecting the pollution control device on their cars to decrease gasoline consumption (some of my students confess to doing this) – but the long-term interests of the whole community are hurt by the polluting of the air. In fact, the lungs of the individual doing the polluting may be hurt in the long run by his or her auto's pollution. While this is true, the attractiveness of the short-term benefit over the long-term interests for any one individual can be overwhelming. Such was the case when I bought a car in 1979. I had the choice of buying a 1979 model, which used leaded gasoline, or a 1980 model, which used unleaded gasoline. Although I knew in a general way at that time that using unleaded gasoline was better for the environment, I bought the 1979 model because leaded gasoline was cheaper.

A way to deal with situations in which individuals gain benefits from polluting, is to use political solutions – solutions designed by the community or its representatives, and which all members of the community will have to obey. In the example involving my students, a possible solution would be more effective auto inspections and the use of steep fines for removing the pollution control device. In the example involving me, a simple solution would have been for the government to place a higher tax on leaded gasoline than on nonleaded gasoline to equalize their price so that there would not have been a monetary advantage for me to pick the 1979 model over the 1980 model.

The exporting of nuclear technology is also likely to be a tragedy-of-the-commons situation. Nuclear reprocessing plants and uranium enrichment plants produce plutonium and uranium, which can be used for making nuclear weapons. The United States, West Germany, and France have competed with each other to sell to Third World countries nuclear reactors and nuclear technologies. The former Soviet Union had a much better record in this area than Western nations, as it was very reluctant to export its nuclear technology. The fact that many of the Third World nations which are getting nuclear technology are politically unstable and involved in heated regional conflicts makes this potential tragedy-of-the-commons situation very dangerous indeed. As more nations acquire nuclear weapons, the odds go up that they will be used. And the danger that a local conflict will draw in other countries is real.[3]

Unanticipated Consequences of the Use of Technology

Ecology is the study of the relationships between organisms and their environments. Without a knowledge of ecology, we are tempted to use technology to solve a single problem. But there are many examples to illustrate the truth that we cannot change one part of the human environment without in some way affecting other parts. Often these other effects are harmful, and often they are completely unanticipated,[4] as the box about cats nicely illustrates.

The Case of the Parachuting Cats

The following situation, which occurred in Borneo, illustrates the unanticipated consequences of the use of technology. There, the efforts of health officials to destroy malaria-carrying mosquitoes by spraying houses with DDT led to the collapsing of the roofs of village houses and to the need to parachute cats into the villages:

[Shortly after the spraying] the roofs of the natives' houses began to fall because they were being eaten by caterpillars, which, because of their particular habits, had not absorbed very much of the DDT themselves. A certain predatory wasp, however, which had been keeping the caterpillars under control, had been killed off in large numbers by the DDT. But the story doesn't end here, because they brought the spraying indoors to control houseflies. Up to that time, the control of houseflies was largely the job of a little lizard, the gecko, that inhabits houses. Well, the geckos continued their job of eating flies, now heavily dosed with DDT, and the geckos began to die. Then the geckos were eaten by house cats. The poor house cats at the end of this food chain had concentrated this material, and they began to die. And they died in such numbers that rats began to invade the houses and consume the food. But, more important, the rats were potential plague carriers. This situation became so alarming that they finally resorted to parachuting fresh cats into Borneo to try to restore the balance of populations that the people, trigger happy with the spray guns, had destroyed.

Source: "Ecology: The New Great Chain of Being," *Natural History,* 77 (December 1968), p. 8.

The use of DDT in the United States has also had major unanticipated effects since it is persistent (it does not easily break down into harmless substances) and poisonous to many forms of life.[5] According to one study, many of the effects of the use of DDT could not have been predicted before its use.[6] The author of this study believes that only through the close monitoring of the effects of new chemicals and through an open debate on those effects can chemicals such as DDT be controlled.

Let's look at factory farms and the unanticipated consequences that have come with the adoption of factory techniques to produce animals for human consumption. Such techniques have been adopted to raise poultry, pigs, veal calves, and cattle. The techniques allow large numbers of animals to be raised in a relatively small space. (Many of these animals never see the light of day until they are removed for slaughter.) The crowding of many animals in a small space and the confinement of individual animals in small stalls creates stress, frustration, and boredom in the animals. Stress can lower the natural defenses of the animals to diseases, and the crowded conditions facilitate the rapid spreading of diseases among the animals. It is common for factory-raised animals to receive large doses of antibiotics in their feed to prevent the outbreak of diseases. There is now evidence that the abundant use of antibiotics in animal food is creating bacteria that are resistant to treatment by modern drugs and that these bacteria can cause illness in humans.[7] Other drugs (to promote rapid growth, for example) and pesticides (to control the highly unsanitary conditions caused by many animals being kept in a small space) are showing up in the meat and poultry coming out of the factory farms. A Department of Agriculture study undertaken between 1974 and 1976 showed that about 15 percent of the meat and poultry it sampled had illegally high levels of drugs and pesticides.[8] But by the late 1980s the situation had improved in the United States because farmers reacted to the increased concern of consumers regarding the drug residues and changed some of their practices and because of increased government enforcement of regulations concerning the use of drugs. Inspections of US slaughterhouses in 1989 revealed that residues of toxic drugs had virtually disappeared from beef and poultry but that some pork and veal still had illegal levels of drug residues.[9]

The Green Revolution had unanticipated negative consequences along with its success in raising grain production in the Third World. In parts of India where the land was relatively evenly distributed before the Green Revolution, the new farming techniques worked well to both increase production and rural employment. But in other parts of the world where the ownership of the land was highly uneven, with a few large landowners and many small ones – a common situation in the Third World – the Green Revolution caused the few rich farmers to get richer and the many poor farmers to get poorer. Here is why that happened:

Large farmers generally adopt the new methods first. They have the capital to do so and can afford to take the risk. Although the new seed varieties do not require tractor mechanization, they provide much

189

economic incentive for mechanization, especially where multiple cropping requires a quick harvest and replanting. On large farms, simple economic considerations lead almost inevitably to the use of labor-displacing machinery and to the purchase of still more land. The ultimate effects of this... are agricultural unemployment, increased migration to the city, and perhaps even increased malnutrition, since the poor and unemployed do not have the means to buy the newly produced food.[10]

The unanticipated consequences of the use of technology can be seen in a situation of which I have some personal knowledge. When I was in Iran in the late 1950s with the US foreign aid program, one of our projects was to modernize the police force of the monarch, the Shah of Iran. We gave the national police new communications equipment so that police messages could be sent throughout the country quickly and efficiently. The United States gave such assistance to the Shah to bolster his regime and help him to maintain public order in Iran while development programs were being initiated. All fine and good, except for the fact that the Shah used his now efficient police – and especially his secret police, which the US CIA helped train – not just to catch criminals and those who were trying to violently overthrow his government, but to suppress all opponents of his regime. His secret police, SAVAK, soon earned a worldwide reputation for being very efficient – and ruthless. Such ruthlessness, which often involved torturing suspected opponents of the Shah, was one of the reasons why the Shah became very unpopular in Iran and was eventually overthrown in 1979 by the Ayatollah Khomeini, a person who had deep anti-American feelings.[11]

Inappropriate Uses of Technology

In 1973, E. F. Schumacher published his book *Small Is Beautiful: Economics as if People Mattered*.[12] This book became the foundation for a movement that seeks to use technology in ways that are not harmful to people. Schumacher argued that the developing nations need intermediate (or "appropriate") technology, not the high (or "hard") technology of the Western industrialized nations. Intermediate technology lies in between the ineffective and primitive technology common in the rural areas of the Third World – where most of the world's people live – and the technology of the industrialized world, which tends to use vast amounts of energy, pollutes the environment, requires imported resources, and often alienates the workers from their own work. The intermediate technology movement seeks to identify those areas of life in the Third World, and also in the industrialized West, where a relatively simple technology can make people's work easier while remaining meaningful, that is, giving them a feeling of satisfaction when they do it.

It is this sense of satisfaction, or contentment, which is often absent in workers in developed nations. A good example of this can be seen in the "workers' revolt" which took place in the ultramodern automobile plant in Lordstown, Ohio, which was to produce Vegas for General Motors and which incorporated the latest in automated technology. The revolt led to a vote by 97 percent of the workers to strike over working conditions. The workers' discontent with the new plant and its mass production techniques can be summed up by the suggestion of one of the strikers that the workers ought to take a sign that was attached to some of the machines, "Treat Me with Respect and I will give you Top Quality Work with Less Effort," and print it on *their* T-shirts.[13]

The high technology of the West is often very expensive, and thus large amounts of capital are needed to acquire it, capital that most Third World nations do not have. This technology is referred to as being capital-intensive instead of labor-intensive. This means that money – but not many people – is needed to obtain it and maintain it. In other words, high technology does not give many workers jobs. (This is the essence of the mass production line: lots of products by a relatively small number of workers.) But the main problem in nations that are trying to develop – and, in fact, in the United States also when its economy is in a recession – is that there are not enough jobs for people in the first place. It is the absence of jobs in the rural areas that is causing large numbers of the rural poor in the Third World to migrate to the cities looking for work, work that is often not there.

While it is fairly obvious, and widely recognized, that developing nations should select technologies that are appropriate to their needs, why don't they always do this? Why has this seemingly simple "lesson" not been learned? The authors of a study of World Bank experiences over nearly four decades explain why they believe inappropriate technology is frequently chosen:

Why does this happen? Foreign consultants or advisers may advocate the technology with which they are most familiar. Local engineers, if educated abroad or the heirs of a colonial legacy, may have acquired a similar bias in favor of advanced technology, or they may simply presume, as do their superiors, that what is modern is best. Special interest groups may favor a particular technical approach.... Deep-seated customs and traditions may favor certain solutions and make others unacceptable. Economic policies that overprice labor (through minimum wage or other legislation) or underprice capital (through subsidized interest rates or overpriced currency) may send distorted signals to decision makers. A simple lack of knowledge or reluctance to experiment may limit the range of choice....When aid is tied to the supply of equipment from the donor country....freedom to choose an appropriate technology may be compromised. With so many factors at work, it is not surprising that a "simple" lesson – such as selecting an appropriate technology – may prove far from simple to apply.[14]

191

I have witnessed the inappropriate use of high technology in both Liberia and the United States. As part of US economic assistance to Liberia, we gave the Liberians road-building equipment. That equipment included power saws. As I proceeded to turn some of this equipment over to Liberians in a small town in a rural area, I realized that the power saws we were giving them were very inappropriate. To people who had little or no experience with power tools – which applied to nearly all the Liberians in that town – the power saw was a deadly instrument. Also, they would not be able to maintain them or repair them when they broke down. Their noise would ruin the peacefulness of the area. A much more appropriate form of assistance would have been crates of axes and hand saws, tools that they could easily learn to use safely, that they would be able to maintain and repair themselves, and that would have provided work for many people.

In the United States I became aware of the inappropriate use of high technology as my wife and I began to prepare for the birth of our children. Most children in the world are born at home, but in the United States and in many other developed countries nearly all births take place in hospitals. An impressive number of studies now show that moving births into hospitals has resulted in unnecessary interventions in the birth process by doctors and hospital staff, which upset the natural stages of labor and can jeopardize the health of both the mother and the baby.[15] As many as 85 to 90 percent of women can give birth naturally, without the use of technology being required.[16] Prenatal care can usually identify the 10 to 15 percent that cannot deliver normally, and for them the use of technology can help protect the lives of the mother and baby. But the major error that has been made is that procedures that are appropriate for these few are now routinely used for most births.

The intermediate technology movement is not against high technology as such (it recognizes areas where high technology is desirable – there is no other way to produce vaccines against deadly diseases, for example), but only against the use of such technology where simpler technology would be appropriate.

Limits to the "Technological Fix"

In our society, which makes wide use of technology, there is a common belief that technology can solve our most urgent problems. It is even believed that the problems that science and technology have created can be solved by more science and technology. What is lacking, according to this way of thinking, is an adequate use of science and technology to solve the problem at hand. In other words, we must find a "technological fix."

While the ability of technology to solve certain problems is impressive, there are a number of serious problems confronting humans – in fact, probably the most serious problems which humans have ever faced

– which seem to have no technological solution. Technology itself has often played a major role in causing these problems. Let's look at a few of them.

The population explosion appears to have no acceptable technological solution. Birth control devices can certainly help in controlling population growth; without such devices a solution to the problem would be even more difficult than it is. But as we have seen in chapter 2, the reasons for the population explosion are much more complicated than the lack of birth control devices. Economic, social, and political factors play a significant role in this situation and must be taken into consideration in any effort to control the explosion. A technological advancement was one of the causes of the population explosion – the wiping out of major diseases, such as smallpox, which used to kill millions. Some people, such as Garrett Hardin, have argued that many of those people who are advocating technological solutions to the population problem such as farming the seas, developing new strains of wheat, or creating space colonies, "are trying to find a way to avoid the evils of overpopulation without relinquishing any of the privileges they now enjoy."[17]

Huge municipal sanitation plants were once considered the solution to our polluted streams, rivers, and lakes, but the rising costs of these plants and the fact that they treat only part of the polluted water are bringing this solution into question.[18] As much water pollution is caused by agricultural and urban runoffs, both of which are not treated by the plants, as by sewage. To talk about a technological fix for this problem is to talk about spending astronomical sums of money to treat all polluted water, and even then the solution would still be in doubt.

A final example will be given to illustrate the limits to the technological fix. As we shall see in the case study below, the nuclear arms race between the Soviet Union and the United States after World War II threatened the world with a holocaust beyond comprehension. Many believed that technology would solve this problem; all that was needed to gain security was better weapons and more weapons than the other side. But the history of the arms race that lasted nearly half a century until the disintegration of the Soviet Union in the early 1990s, clearly shows that one side's advantage was soon matched or surpassed by new weapons on the other side. Momentary feelings of security by one nation were soon replaced by deepening insecurity felt by both nations as the weapons became more lethal. "Security dilemma" is the phrase that has been coined to describe a situation where one nation's efforts to gain security lead to its opponent's feeling of insecurity. This insecurity causes the nation that believes it is behind in the arms race to build up its arms, but it also causes the other nation to feel insecure. So the race goes on. The temptation to believe that a new weapon will solve the problem is immense. A brief history of the arms race shows how both superpowers were caught in a security dilemma.

The United States exploded its first atomic bomb in 1945 and felt fairly secure until the Soviets exploded one in 1949. In 1954 the United

193

States tested the first operational thermonuclear weapon (a hydrogen or H-bomb), which uses the A-bomb as a trigger, and a year later the Soviets followed suit. In 1957 the Soviets successfully tested the first intercontinental ballistic missile (ICBM) and launched the earth's first artificial satellite, Sputnik. The United States felt very insecure, but within three years had more operational ICBMs than the Soviet Union.[19]

The Soviet Union put up the first antiballistic missile system around a city – around Moscow – in the 1960s, and in 1968 the United States countered by developing MIRVs (multiple, independently targetable reentry vehicles), which can easily overwhelm the Soviet antiballistic missiles. The Soviets started deploying their first MIRVs in 1975, and these highly accurate missiles with as many as ten warheads on a single missile, each one able to hit a different target, led President Reagan in 1981 to declare that a "window of vulnerability" existed, since the land-based US ICBMs could now be attacked by the Soviet MIRVs. Reagan began a massive military buildup.

The technological race was poised to move into space when President Reagan in 1983 announced plans to develop a defensive system, some of which would probably be based in space, which could attack any Soviet missiles fired at the United States. This system (formally known as the Strategic Defense Initiative, and informally called "Star Wars"), was criticized by many US scientists as being infeasible and by the early 1990s had been greatly reduced in scope.

An unexpected end to the nuclear arms race between the Soviet Union and the United States came in the late 1980s with the collapse of the Soviet empire in Europe and with the breakup of the Soviet Union itself in the early 1990s. The huge financial strain on its economy caused by the arms race undoubtedly contributed to its collapse. But the nuclear arms race also placed serious strains on the US economy.[20] The end of the Cold War brought a nearly miraculous release to the world from the danger of a third world war, which likely would have been the world's last one. But nuclear weapons continue to exist, and as we will see in the case study below, they still represent a great threat to life on earth.

War ●

Why do human beings make war? Some of the people who have studied the causes of war believe that war is caused by the negative aspects of human nature, such as selfishness, possessiveness, irrationality, and aggressiveness. Other students of war have come to the conclusion that certain types of government – or, more formally, how political power is distributed within the state – make some countries more warlike than others. And other analysts have concluded that international anarchy, or the absence of a world government where disputes can be settled peacefully and authoritatively, is the main cause of war. Kenneth Waltz, a respected US student of war, has concluded that human nature, and/or the type of government are often the immediate causes of war, but that international anarchy explains why war has recurred throughout human history. [21]

War reflects the relatively primitive state of human political development. When Albert Einstein, the German-born American theoretical physicist who is considered to be one of the most brilliant persons of the twentieth century, was reportedly asked why is it that we are able to create nuclear weapons but not abolish war, he responded that the answer was easy: politics is more difficult than physics.

The nations of the world spend about $600 billion a year on preparing for war. Since the end of the Cold War, there has been a trend to reduce the funds spent on military forces, but the huge amount still spent on the military leads to disturbing comparisons. Here are just two. The world spends about 25 times more money supporting each soldier as it does each child enrolled in school. There are six times as many soldiers as there are physicians in the Third World.

Since World War II there have been about 150 wars, with 90 percent of those occurring in the less developed nations. Wars have been frequent in the Third World since 1945 for a number of reasons. During the Cold War the United States and the Soviet Union supported with arms various political groups in the less developed nations that favored their side in the East/West conflict. Although the Cold War has now ended, the huge amounts of weapons supplied by the superpowers are now circulating within the Third World (and even in conflicts in Europe such as in the former Yugoslavia). Conflicts have been frequent in the Third World also because many of these nations received political independence relatively recently and territorial disputes, power struggles, ethnic and religious rivalries, and rebellions caused by unjust conditions are common. The military controls political power in about one-half of the developing nations, and these regimes are usually involved in more military conflicts than are civilian-controlled governments.[22]

A characteristic of modern war is that often more civilians are killed than soldiers. From 1945 to 1992 an estimated 7.5 million military personnel died in war while about twice that number (14.5 million) civilians were killed in wars.[23] In many wars in the past the military combatants were the main casualties but this has now changed so that civilians often bear the greatest burden. If one adds to the number of civilians killed and wounded during the fighting the vast number of civilians who flee the fighting and become refugees – sometimes finding no place which will accept them – civilians indeed bear the largest burden of modern war. Also the destruction of the land by the fighting is often immense so that when the fighting finally ends, civilians return to an ecologically damaged land.

Another characteristic of modern wars is that technology has been used to greatly increase the destructive capacity of the weapons. The case study on nuclear weapons that follows this section will illustrate that point well, but even so-called conventional weapons are now much more destructive than they used to be. (See the box on modern, high-tech weapons.) In addition to the increase in destructive capacity, technology has been used to increase the weapons' accuracy, penetration ability, rates of fire, range , automation, and armor.

Modern High-Tech Weapons

The following is a description by Paul Walker, a military specialist, of some of the new weapons used by the United States during the 1991 war to push Iraq out of Kuwait:

The BLU-97/B is a three-quarter pound bomblet which carries a triple punch: a pre-fragmented anti-personnel casing to spray deadly shrapnel; a hollow-charge anti-tank warhead; and a disc of incendiary zirconium. Whatever is left after the shrapnel and bomb is lapped up by fire.

The laser-guided bombs which destroyed the air-raid shelter in Iraq, a refuge for over 500 Iraqi civilians, had high penetration noses with delayed fuses mounted on the tails so as to explode only after entering hardened targets like the bunker.

The CBU-87B is a 950-pound bomb which carries 202 small bomblets. One B-52 plane loaded with the CBU-87B's can carpet-bomb over 176 million square yards, equal to 27,500 football fields.

The MADFAE (mass air delivery fuel-air explosive) mimics small nuclear explosions. It consists of 12 containers of ethylene oxide or propylene oxide. Trailed behind utility helicopters, the containers release a cloud of highly volatile vapors which, when mixed with air and detonated, can cover an area over 1,000 feet long with blast pressures five times that of TNT.

Source: Ruth L. Sivard. *World Military and Social Expenditures 1993*, 15th edn (Washington: World Priorities, 1993), p. 18.

There are some positive signs regarding war. The world appears to be making some progress in controlling war. Here is an assessment of our present situation by Ruth Sivard, the author of a respected review of military and social conditions in the world:

Despite the evident disorder of the contemporary world, there are now a number of signs that basic trends - not only of public opinion but also of public policy - are in directions favorable to peace and social development. Foremost is the radical change in superpower relations. Overall, the arms race is moderating. The United Nations is taking a much more active role in peacekeeping and peacemaking. International agreements have been signed that will reduce arms, not merely control them. Major powers are cutting back on weapons purchases and inventories, as well as on the number of men and women in the armed forces. Although the pace of reduction is still small, world military expenditures have declined for the fifth successive year. The omens at last look good for a needed post–Cold War dividend.[24]

The Threat of Nuclear Weapons: A Case Study

The threat of nuclear weapons is a subject that touches on many of the themes we have examined in this chapter. It is the "ultimate" development subject since it is the achievements of weapons technology by the developed nations that have brought the survival of human life into question. It is a problem that cries out for a political solution. Karl von Clausewitz, the famous Prussian author of books on military strategy, described war as a continuation of politics by other means. But, given the probable consequences of a nuclear war as presented below, one must ask whether war between nations with nuclear weapons can remain a way of settling their disputes? Let us look at the nature of the threat that nuclear weapons have created and then at three contemporary problems caused by these weapons.

The Threat

It has taken 4.5 billion years for life to reach its present state of development on this planet. The year 1945 represents a milestone in that evolution since it was then that the United States exploded its first atomic bombs on Hiroshima and Nagasaki, Japan, and demonstrated that humans had learned how to harness for war the essential forces of the universe. After 1945, when the United States had no more than two or three atomic bombs, the arms race continued until the two superpowers, the United States and the Soviet Union, had a total of about 50,000 nuclear weapons, the equivalent of 1 million Hiroshima bombs – or, to put it another way, about three tons of TNT for every man, woman, and child in the world. The Hiroshima bomb was a 15 kiloton device (a kiloton having the explosive force of 1,000 tons of TNT); some of the weapons today fall in the megaton range (a megaton being the equivalent of 1 million tons of TNT[25]). Today, in addition to the United States and the Commonwealth of Independent States (created from most of the republics in the Soviet Union), Britain, France, and China have nuclear weapons that could be used in a nuclear war. The United States and the Commonwealth of Independent States have agreed to greatly reduce their strategic nuclear stockpiles by the beginning of the next century (to 3,000–3,500 warheads each), but the implementation of these agreements is still in question, given the political and economic instability in the former Soviet republics where the weapons are located.

What would happen if these weapons were ever used? We cannot be sure of all the effects, of course, since, as the author Jonathan Schell has stated, we have only one earth and cannot experiment with it.[26] But we do know from the Hiroshima and Nagasaki bombings, and from the numerous testings of nuclear weapons both above and below ground, that there are five immediate destructive effects from a nuclear explosion: (1) the initial radiation, mainly gamma rays; (2) an electromagnetic pulse, which in a high-altitude explosion can knock out electrical equipment over a very large area; (3) a thermal pulse, which consists of bright light (you would be blinded by glancing at the fireball even if you were many miles away) and intense heat (equal to that at the center of the sun); (4) a blast wave that can flatten buildings; and (5) radioactive fallout, mainly in dirt and debris that is sucked up into the mushroom cloud and then falls to earth.

The longer-term effects from a nuclear explosion are three: (1) delayed or worldwide radioactive fallout, which gradually over months and even years falls to the ground, often in rain; (2) a change in the climate (possibly a lowering of the earth's temperature over the whole Northern Hemisphere, which could ruin agricultural crops and cause widespread famine); and (3) a partial destruction of the ozone layer, which protects the earth from the sun's harmful ultraviolet rays. If the ozone layer is depleted, unprotected Caucasians could stay outdoors for only about ten minutes before getting an incapacitating sunburn (blacks, because of the color of their skin, could go somewhat longer), and people would suffer a type of snow blindness from the rays which, if repeated, would lead to permanent blindness. Many animals would suffer the same fate.[27]

Civil defense measures might save some people in a limited nuclear war but would not help much if there were a full-scale nuclear war.[28] Underground shelters in cities hit by nuclear weapons would be turned into ovens since they would tend to concentrate the heat released from the blast and the firestorms. Nor does evacuation of the cities look like a hopeful remedy in a full-scale nuclear war, since people would not be protected from fallout, or from retargeted missiles, and could not survive well in an economy that had collapsed.

Since most of our hospitals and many doctors are in central-city areas and would be hit by the first missiles in an all-out nuclear war, medical care would not be available for the millions of people suffering from burns, puncture wounds, shock, and radiation sickness. Many corpses would remain unburied and would create a serious health hazard, which would contribute to the danger of epidemics spreading among the population whose resistance to disease had been lowered by radiation exposure, malnutrition, and shock.

What could be the final result of all of this? Here is how Jonathan Schell answers that question in probably the longest sentence you have ever read, but in one with no wasted words:

> Bearing in mind that the possible consequences of the detonations of thousands of megatons of nuclear explosives include the blinding of insects, birds, and beasts all over the world; the extinction of many ocean species, among them some at the base of the food chain; the temporary or permanent alteration of the climate of the globe, with the outside chance of "dramatic" and "major" alterations in the structure of the atmosphere; the pollution of the whole ecosphere

with oxides of nitrogen; the incapacitation in ten minutes of unprotected people who go out into the sunlight; the blinding of people who go out into the sunlight; a significant decrease in photosynthesis in plants around the world; the scalding and killing of many crops; the increase in rates of cancer and mutation around the world, but especially in the targeted zones, and the attendant risk of global epidemics; the possible poisoning of all vertebrates by sharply increased levels of vitamin D in their skin as a result of increased ultraviolet light; and the outright slaughter on all targeted continents of most human beings and other living things by the initial nuclear radiation, the fireballs, the thermal pulses, the blast waves, the mass fires, and the fallout from the explosions; and considering that these consequences will all interact with one another in unguessable ways and, furthermore, are in all likelihood an incomplete list, which will be added to as our knowledge of the earth increases, one must conclude that a full-scale nuclear holocaust could lead to the extinction of mankind.[29]

Underground nuclear weapons testing site in the USA. (*Los Alamos National Laboratory*)

New Dangers

Despite the end of the Cold War and of the threat of a cataclysmic war between two superpowers, nuclear weapons still remain a danger for the world. Three problems exist with which the world will have to deal: (1) the proliferation of nuclear powers; (2) the control of the nuclear weapons remaining in the new countries created from the breakup of the Soviet Union; and (3) the cleanup of the huge amount of toxic wastes produced in both the United States and the former Soviet Union when they built their large numbers of nuclear weapons.

Nuclear Proliferation

The spread of nuclear weapons to new countries represents a growing danger because the larger the number of countries that have these weapons the greater the likelihood that they will be used. Many of these new nuclear powers – either actual or potential – are Third World authoritarian regimes that have serious conflicts with their neighbors. For example, the Middle East is a region plagued by conflict. It is widely believed that Israel has already acquired nuclear weapons and has them ready for use or could have them ready in a very short time. After the defeat of Iraq in the Persian Gulf War in 1991, UN inspectors discovered that Iraq had been making major efforts to build both atomic weapons as well as the much more powerful hydrogen weapons. This was in spite of the fact that Iraq had signed the Nuclear Nonproliferation Treaty in which it had agreed not to acquire nuclear weapons and in spite of the fact that officials from the International Atomic Energy Agency had inspected nuclear facilities in Iraq just prior to the war and had found no evidence that Iraq was building nuclear weapons.

Another example of proliferation is in South Asia. In this region two countries – India and Pakistan – have already fought each other three times in the past 40 years and both are believed to have built nuclear weapons.

Regional conflicts in which these weapons could be used are not the only concern; also disturbing is the possibility of accidental or unauthorized use of nuclear weapons by these countries.[30] In the early 1990s the list of actual and potential nuclear powers was as follows:

Acknowledged Nuclear Powers: United States, Russia, Ukraine, Kazakhstan, Belarus (last four newly independent states created from the former Soviet Union), United Kingdom, France, China

Suspected Nuclear Powers: Israel, India, Pakistan, South Africa

Past and Present Suspected Aspiring Nuclear Powers: Algeria, Argentina, Brazil, Iran, Iraq, Libya, North Korea, South Korea, Taiwan

Control of Nuclear Weapons in the former Soviet Union

With the breakup of the Soviet Union there is a concern about who will control the thousands of nuclear weapons that still exist there. There is a general agreement among the members of the Commonwealth of Independent States, that only Russia should retain nuclear weapons. But the implementation of this agreement is still in doubt. And even if Russia does become the only nuclear state, who will control the weapons there? With the collapse of communism has come frightening political, economic, and social instability in the former Soviet republics. Until these states obtain a new stability, it is disquieting, to say the least, to remember that many nuclear weapons remain in this area.

Also of concern is the fate of the thousands of former Soviet scientists who possess experience in the building of nuclear weapons. As the economy in the newly independent states deteriorates, where will these people find employment? Will they be tempted by possible foreign offers of jobs by some of the aspiring nuclear states listed above? Or by terrorists? Obviously these are not pleasant possibilities but ones the world can not safely fail to consider.

The Cleanup

The production of vast quantities of nuclear weapons in both the Soviet Union and in the United States led to huge environmental contamination with highly toxic chemical and nuclear wastes. In both countries wastes from the plants producing components for the nuclear weapons were released into the air and dumped onto the ground, and they have leaked from temporary storage facilities. The extent of this contamination did not become public until the late 1980s in the US case when the US government released a number of reports outlining the huge extent of the problem and, in the Soviet case, in the late 1980s and the early 1990s in the last years of the Soviet communist state.[31]

It is painful to read about the deliberate inflicting of harm by a government on its own citizens. Although the Soviet contamination is probably greater than the American,[32] both governments used "national security" to justify their actions and to keep them secret. In the United States the plants were exempt from state and federal environmental laws, and actions were carried out that had long before been declared illegal for private industry and individuals. An estimated 60,000 nuclear weapons were made in the United States over a 45-year period, in 15 major plants covering an area equal in size to the state of Connecticut. They cost about $300 billion (in 1991 dollars). Estimates by various government agencies of the cost of cleaning up the environmental damage at the plants, which will take decades to accomplish, range from $100 billion to $300 billion.[33]

201

Conclusions

This chapter has focused on the negative aspects of technology. It has done so because most of the readers of this book will probably be citizens of developed countries who already have a strong belief in the advantages of technology. It is not my intent to undermine that belief, because technology has benefited human beings in countless ways, and its use is largely responsible for the high living standards in the United States and other industrialized nations. Rather, my intent is to bring a healthy caution to the use of technology. An ignoring of the negative potential of technology has brought harm to people in the past and could cause unprecedented harm in the future. Much technology is neither good nor bad. It is the use that human beings make of this technology that determines whether it is mainly beneficial or harmful. But other technology is basically harmful or excessively dangerous and should be rejected. It is of course not always easy to place technologies in these categories, but an effort should be made.

The less developed nations need technology to help them solve many of their awesome problems. But often intermediate technology should be used by them rather than the high technology of the industrialized nations. The temptation to imitate the West is strong, but ample evidence exists to show that this could be a serious mistake for developing nations. The Third World needs to remember that its conditions and needs are different from those of the West, and that it should take from Western science only what is appropriate.

The industrial nations face another task. They must become more discriminating in their use of technology and lose some of their fascination with and childlike faith in it. The fate of the earth is now literally in their hands, especially in those of the United States and the former Soviet Union. The wisdom or lack of wisdom these nations show in using military technology affects all – the present inhabitants of earth, both human and nonhuman, and future generations, who depend on our good judgment for their chance to experience life on this planet.

Notes

1 Garrett Hardin, "The Tragedy of the Commons," *Science*, 162 (December 13, 1968), pp. 1243–8.

2 Donald J. Puchala, "American Interests and the United Nations," *Political Science Quarterly*, 97 (Winter 1982–3), p. 585.

3 See Nigel Calder, *Nuclear Nightmares: An Investigation into Possible Wars* (New York: Penguin Books, 1979), ch. 3, "The Nuclear Epidemic," on how this can happen.

4 A description of 50 case studies of development projects in the Third World that had harmful and unanticipated effects on the environment is contained in the following conference report: M. Taghi Farvar and John P. Milton, eds, *The Careless Technology: Ecology and International Development* (Garden City, NY: Natural History Press, 1972).

5 Although its use was banned in the United States in 1972, residues of

DDT could still be found in most Americans 20 years later. DDT is stored in the human body for decades."Study Links DDT and Cancer," *New York Times*, national edn (April 22, 1993), p. A10.

6 Thomas R. Dunlap, *DDT: Scientists, Citizens, and Public Policy* (Princeton: Princeton University Press, 1981), p. 8.

7 *New York Times*, national edn (July 1, 1982), p. 10.

8 Jim Mason and Peter Singer, *Animal Factories* (New York: Crown, 1980), p. 53. For the effects on the consumers, the farmers, and the animals themselves of using factory methods to raise animals for human consumption, see also Peter Singer, *Animal Liberation: A New Ethics for Our Treatment of Animals* (New York: New York Review, 1975), ch. 3.

9 Keith Schneider, "Toxic Drug Residues in Meat: U.S. Says Findings Are Mixed," *New York Times*, national edn (June 1, 1990), p. A11.

10 Donella H. Meadows et al., *The Limits to Growth*, 2nd edn (New York: Universe Books, 1974), p. 147.

11 For a fuller discussion of the unanticipated consequences of American aid to the Shah, see John L. Seitz, "The Failure of U.S. Technical Assistance in Public Administration: The Iranian Case," *Public Administration Review*, 40 (September-October 1980), pp. 407-13.

12 E. F. Schumacher, *Small Is Beautiful: Economics as if People Mattered* (New York: Harper and Row, 1973).

13 Emma Rothschild, *Paradise Lost: The Decline of the Auto-Industrial Age* (New York: Random House, 1973), p. 119.

14 Warren C. Baum and Stokes M. Tolbert, *Investing in Development: Lessons of World Bank Experience* (Oxford: Oxford University Press, 1985), p. 574.

15 See, for example, Suzanne Arms, *Immaculate Deception: A New Look at Women and Childbirth in America* (Westport, Conn.: Bergin and Garvey, 1984); Dr Robert A. Bradley, *Husband-coached Childbirth* (New York: Harper and Row, 1974); Robbie E. Davis-Floyd, *Birth as an American Rite of Passage* (Berkeley: University of California Press, 1992); Barbara K. Rothman, "Midwives in Transition: The Structure of a Clinical Revolution," *Social Problems*, 30 (February 1983), pp. 262–71; and Neal Devitt, "The Transition from Home to Hospital Birth in the United States, 1930-1960," *Birth and the Family Journal*, 4 (Summer 1977), pp. 47-58.

16 Dr John S. Miller, "Foreword." In Lester D. Hazell, *Commonsense Child-birth* (New York: Berkley Books, 1976), p. x.

17 Hardin, "The Tragedy of the Commons," p. 1243.

18 Jon R. Luoma, "The $33 Billion Misunderstanding," *Audubon*, 83 (November 1981), pp. 111-27.

19 This "missile gap," in which the Soviets trailed, could have been the reason they put missiles in Cuba in 1962, which led to the Cuban missile crisis, the world's first approach to the brink of nuclear war. The humiliation the Soviet Union suffered when it had to take its missiles out of Cuba may have led to its build-up of nuclear arms in the 1970s and 1980s, which caused great concern in the United States.

20 The economies of other countries such as Germany and Japan grew stronger while the United States and the Soviet Union were involved in the arms race so that some recognized the truth in the joke told in the early 1990s that "The Cold War between the United States and the Soviet Union is over and Japan is the winner."

21 Kenneth A. Waltz, *Man, the State and War: A Theoretical Analysis* (New York: Columbia University Press, 1959).

22 Ruth L. Sivard, *World Military and Social Expenditures, 1993*, 15th edn (Washington: World Priorities, 1993), p. 22.

23 Ibid., p. 21.

24 Ibid., p. 7.

25 A train transporting 1 million tons of TNT would be about 250 miles long.

26 Jonathan Schell, *The Fate of the Earth* (New York: Avon Books, 1982).

27 For a fuller description of the effects of a nuclear war see Schell, *The Fate of the Earth*, ch. 1; Ruth Adams and Susan Cullen (eds), *The Final Epidemic: Physicians and Scientists on Nuclear War* (Chicago: Bulletin of the Atomic Scientists, 1981); and "Nuclear War: The Aftermath," *Ambio: A Journal of the Human Environment* (Royal Swedish Academy of Sciences, Pergamon Press) 11/2-3, (1982).

28 For an interesting discussion of the negative American attitude toward civil defense, see Freeman Dyson, *Weapons and Hope* (New York: Harper & Row, 1984), especially ch. 8.

29 Schell, *The Fate of the Earth*, p. 93.

30 Good discussions of these problems are contained in Bruce G. Blair and Henry W. Kendall, "Accidental Nuclear War," *Scientific American* 263 (December 1990), pp. 53–8, and in William Broad, "Guarding the Bomb: A Perfect Record, but Can It Last?" *New York Times*, national edn (January 29, 1991), pp. B5, B8. For further information about the nuclear proliferation danger see, John M. Deutch, "The New Nuclear Threat," *Foreign Affairs* 71 (Fall 1992), pp. 120–34.

31 For details of the contamination at US plants as described by US government agencies, see the following articles in *New York Times*, national edn: Keith Schneider, "Candor on Nuclear Peril," (October 14, 1988), p.1.; Kenneth Noble, "U.S., For Decades, Let Uranium Leak At Weapons Plant," (October 15, 1988), p. 1; Keith Schneider, "Wide Threat Seen in Contamination At Nuclear Units," (December 7, 1988), p. 1; Matthew Wald, "Waste Dumping That U.S. Banned Went On at Its Own Atom Plants," (December 8, 1988), p. 1; Matthew Wald, "Wider Peril Seen in Nuclear Waste From Bomb Making," (March 28, 1991), p. A1.

32 Matthew Wald, "High Radiation Doses Seen for Soviet Arms Workers," *New York Times*, national edn (August 16, 1990), p. A3.

33 Jim Bencivenga, "Scale-Down of US Nuclear Arsenal Will Affect Widespread Industry," *Christian Science Monitor*, (November 29, 1991), p. 4.

Further Readings

Bass, Thomas, *Camping with the Prince and Other Tales of Science in Africa* (New York: Penguin, 1991). The author, in a "popular" type book, describes people using science and technology in Africa in a way that works for Africa.

Bellini, James, *High-Tech Holocaust* (San Francisco: Sierra Club Books, 1989). Bellini focuses upon the unforeseen damages wrought on the environment by technologically induced disasters supported by greed and deception.

Brennan, Richard P., *Levitating Trains and Kamikaze Genes* (New York: Wiley, 1990). Brennan discusses recent technological breakthroughs in energy, genetic engineering, and biotechnology and explores the benefits as well as the environmental penalties.

Bundy, McGeorge, and William J. Crowe Jr, *Reducing Nuclear Danger. The Road Away From the Brink* (New York: Council on Foreign Relations Press, 1993). A readable, small volume arguing that while the risk of a nuclear world war has greatly diminished in the post–Cold-War world, the overall level of nuclear danger probably has not diminished. The authors make a number of suggestions about how the danger can be reduced.

Del Tredici, Robert, *At Work in the Fields of the Bomb* (New York: Perennial Library/Harper and Row, 1987). Contains impressive photographs of the facilities where the nuclear weapons were built in the United States and interviews with some of the people involved with building the weapons and with those affected by them.

Leiss, William, *Under Technology's Thumb* (Montreal: McGill-Queen's University Press, 1990). Leiss contemplates the effect of technology not only on the environment, but also on the society and its attitude toward the earth and its resources.

Lewis, H. W., *Technological Risk* (New York: W. W. Norton, 1990). Lewis examines the perceived risks associated with our technological society and argues that bad policy, misused resources, and lack of education rather than technology itself are the culprits. He argues that many things people fear the most actually pose no real risk to them, whereas some things that do pose a real risk to people, are not perceived as being dangerous.

Morone, Joseph G., and Edward J. Woodhouse, *Averting Catastrophe: Strategies for Regulating Risky Technologies* (Berkeley: University of California Press, 1986). The authors believe that it is possible to create a risk avoidance system that will make events like Chernobyl and Bhopal unlikely.

Pacey, Arnold, *Technology in World Civilization* (Cambridge: MIT Press, 1990), Pacey explores the history of technology and its diffusion around the world, arguing that a lack of understanding of how this diffusion works has led to misguided programs in developing nations.

Pradervand, Pierre, *Listening to Africa: Developing Africa from the Grassroots* (New York: Praeger, 1989). The author uses case studies to highlight the problems facing Africa and the triumph of "low tech" solutions.

Sivard, Ruth L., *World Military and Social Expenditures 1993*, 15th edn (Washington: World Priorities, 1993). From the title, one would guess that this is a dry, dull book, containing many statistics. In fact, this small volume is filled with interesting text, charts, and figures describing the world's military and social priorities.

Weir, David, *The Bhopal Syndrome* (San Francisco: Sierra Club Books, 1987). An analysis of the Bhopal, India, disaster and examinations of other pesticide plants fuels this argument that the spread of ill-monitored and deadly technology to developing countries exacts tremendous costs to human life, the economy, and the environment.

Winner, Langdon, *The Whale and the Reactor: A Search for Limits in an Age of High Technology* (Chicago: University of Chicago Press, 1986). The sight of a whale surfacing near a nuclear reactor causes the author to contemplate the connections between nature and technology and to call for a more conscious effort by people to think about how technology can affect human life.

7

Alternative Futures

I n human affairs, the logical future, determined by past and present conditions, is less important than the willed future, which is largely brought about by deliberate choices – made by the human free will.

René Dubos (1901–82)

Where is development leading us? What can we say about the future? Probably the wisest thing we can say is that the future is essentially unknowable; it cannot be predicted. If this is so – and the dismal record of past predictions leads us to believe it is – then we might ask, "Does it make any sense to think about the future at all?" I would answer, "Yes, it does." Although the complexity of life and natural and spiritual forces make the future unknowable, human actions can make one future more likely than another. It is this ability to influence the future that concerns us in this final chapter.

If we can accurately recognize some of the major trends and currents in the past and the present, we can make an educated guess about where we are heading. And if we do not like the direction in which the world is heading, we can examine our individual behaviors and governmental policies to determine if they should be changed so that they contribute to a more desired future. As an old Chinese proverb states, if you do not change the direction in which you are headed, you will end up where you are headed.

René Dubos, the late well-known bacteriologist, coined the phrase, "Think globally, act locally." This is the way, according to Dubos, that an individual can help bring about a desirable future. Dubos no doubt realized that the great benefit of local action is that not only does it help solve problems but it helps the individual's spirit grow. It also effectively combats the sense of powerlessness and depression that can come by "thinking globally," by becoming aware of the immense and serious problems the world faces. Wendell Berry, the US author who writes about the need for a connection to the land, gives the following tribute to "local action": "The real work of planet-saving will be small, humble, and humbling, and (insofar as it involves love) pleasing and rewarding. Its jobs will be too many to count, too many to report, too many to be publicly noticed or rewarded, too small to make anyone rich or famous."[1] It is my belief that politics – the process a society uses to achieve commonly desired goals and to settle conflicts among groups with different interests - will also play a central role in determining what the future will be like.

Of the many possible futures the human race faces on earth, three look most likely at present: doom, growth, and sustainable development. Some catastrophe might lead to the death of hundreds of millions of people; economic growth might continue into the future; or the world might achieve development that can continue indefinitely because it does not undermine the environment and resource base upon which it rests. There could, of course, be a combination of two of these or of all three. For example, one part of the world might experience a harsher life in the future while another part continues to expand economically. Or part of the world could reach sustainable development while another part continues to grow. Or even within one country, two or all three scenarios might exist at different times in the future: a period of growth could be followed by a catastrophe that is followed by sustainable development. In the rest of the chapter we will examine what the main proponents of these three views say, and then end with my assessment of the future.

Doom ●

There are a number of writings today warning that if humankind does not change its ways some kind of disaster will occur in the future. The implicit or explicit purpose of most of these authors is to help prevent the expected disaster by suggesting changes in human behavior or policies. Thus these works are not really predictions of what the future will be like, but rather what it could be like if present trends continue. The most frequently discussed disasters are those caused by nuclear weapons, food shortages, pollution, overpopulation and overconsumption, the depletion of nonrenewable resources, and cosmic collisions.

Nuclear Weapons

Jonathan Schell, author of *The Fate of the Earth*[2], is probably the best-known writer who argues that the nation-state system , with its competition among big nations armed with nuclear weapons and with the proliferation of nuclear weapons to Third World nations, is leading the world to a nuclear war. Schell argues, as was pointed out more fully in chapter 6, that such a war could bring about the extinction of human life on our planet. With the end of the Cold War and the collapse of the former Soviet Union, the danger of a full-scale nuclear war has been significantly reduced, but the spread of nuclear arms to new nations allows the nuclear danger to continue. Any use of a nuclear weapon would result in huge casualties and extensive environmental damage.

Pollution, Famines, Overpopulation, Resource Depletion

Since the early 1960s, several widely publicized books have been written that forecast some sort of disaster coming because of industrial pollution, scarcity of food, overpopulation, or depletion of nonrenewable resources. One of the first was Rachel Carson's *Silent Spring*, which predicted premature death to humans and other animals because of the growing use of pesticides and other chemicals.[3] Another book, which received about as much publicity as Carson's book, was *The Population Bomb* by Paul Ehrlich. Ehrlich spelled out the threat of overpopulation in the following terms:

> The battle to feed all of humanity is over. In the 1970s and 1980s hundreds of millions of people will starve to death in spite of any crash programs embarked upon now.... No changes in behavior or technology can save us unless we can achieve control over the size of the human population. The birth rate must be brought into balance with the death rate or mankind will breed itself into oblivion.[4]

Paul Ehrlich, along with his wife, Anne Ehrlich, continued with this theme in their 1990 book, *The Population Explosion*.[5]

Another well known book in the 1970s was *The Limits to Growth*, which was a report by the Club of Rome, a private group concerned with world problems. The report was based on a computer analysis of the world's condition by a research team at the Massachusetts Institute of Technology. The book emphasized that the earth is finite and that there are definite limits to its arable land, nonrenewable resources, and ability to absorb pollution. The study's main conclusion was as follows:

> If the present growth trends in world population, industrialization, pollution, food production, and resource depletion continue unchanged, the limits to growth on this planet will be reached sometime within the next 100 years. The most probable result will be a

rather sudden and uncontrollable decline in both population and industrial capacity.

It is possible to alter these growth trends and to establish a condition of ecological and economic stability that is sustainable far into the future.[6]

In 1992 a sequel to *The Limits to Growth* was published by three of the original authors. This book, *Beyond the Limits: Confronting Global Collapse, Envisioning a Sustainable Future*, still presents the possibility of "overshoot and collapse" if present trends continue, but unlike their earlier book, *Beyond the Limits* focuses more on how collapse can be averted.[7]

In the late 1970s the US government conducted a three-year study of what the world would be like in the year 2000 if present trends continued. The conclusions of the *Global 2000 Report to the President* were consistent with those of the earlier Club of Rome studies:

If present trends continue, the world in 2000 will be more crowded, more polluted, less stable ecologically, and more vulnerable to disruption than the world we live in now. Serious stresses involving population, resources,and environment are clearly visible ahead. Despite greater material output, the world's people will be poorer in many ways than they are today.

For hundreds of millions of the desperately poor, the outlook for food and other necessities of life will be no better. For many it will be worse. Barring revolutionary advances in technology, life for most people on earth will be more precarious in 2000 than it is now–unless the nations of the world act decisively to alter current trends.[8]

The doomsday scenario is clearly seen in a major book of the early 1990s by the Yale University historian Paul Kennedy. *Preparing for the Twenty-First Century* outlines the major challenges the world will have to deal with in the next century: the population explosion, the globalizing of the world's economy and the increasing power of multinational corporations, technological advances in agriculture and in industry (mainly biotechnology and robotics), and environmental damage. Kennedy sees these challenges as having a potentially catastrophic effect on many poor nations, but even the rich ones will be harmed by them if they do not recognize them and come up with imaginative ways to deal with them.[9]

Triage and Lifeboat Ethics

A forecast of doom for the future of humanity has led some authors to recommend policies designed to deal with such situations. One called "triage" was discussed in *Famine-1975!* by William and Paul Paddock.[10] Triage is a procedure that was used in World War I when doctors in battle-

Cosmic Collisions

In 1989 an asteroid, a half-mile in diameter, missed hitting earth by six hours. The asteroid was spotted only after it had already passed earth. After that near miss the US Congress instructed the National Aeronautics and Space Administration (NASA) to study the possibility that an object from space – an asteroid or a comet – could collide with earth and to offer recommendations about how to deal with the danger. NASA's reports were received with a fair degree of skepticism, but that skepticism all but disappeared in the summer of 1994 when astronomers on earth recorded the spectacular fireballs (some as large as earth) on Jupiter as it was hit by the fragments of a comet. Although there is still debate about how likely it is such a collision could take place with earth (the NASA report said the risk of a major collision soon was slim but not negligible), the danger is now being taken seriously. As the spokesman for the American Physical Society, the most prestigious organization of physicists in the United States,

said after the Jupiter event, "Nobody is going to dismiss this."

Some scientists now believe it is possible that a collision with an object from space 65 million years ago led to the extinction of the dinosaurs and many other forms of life as a huge amount of dust rose to block sunlight and drastically lowered the temperature on earth. Since that time, numerous objects have hit earth, some with the force of many nuclear bombs. It is likely that NASA will now recommend a series of telescopes be built around earth to track objects that could hit earth. Some scientists are suggesting that if such an object is discovered early enough, nuclear tipped rockets could be fired at it to attempt to divert it from its collision course.

Sources: William Broad, "Asteroid Defense: Planners Envision Real Possibility of Collision with Earth," *New York Times*, national edn (April 7, 1992), p. B5, and William Broad, "When Worlds Collide: A Threat to the Earth Is a Joke No Longer," *New York Times*, national edn (August 1, 1994), p. A1.

field hospitals had to decide which of the many wounded would receive the limited medical care available. The wounded were divided into three categories. The first were those soldiers who were only slightly wounded and, although in pain, would probably survive even if untreated. The second category consisted of soldiers who were so seriously wounded that even if they received medical attention, they would probably die. In the third category were soldiers who were seriously wounded but who could probably be saved if the doctors treated them. It was to the last category that the military doctors first turned their attention. The Paddock brothers recommended that the United States place the countries of the world who were requesting food aid into three categories similar to those in the triage procedure and give aid only to those countries in the third category, that is, to those which would have a good chance of progressing to a state of being able to survive by their own efforts if they received some aid.

"Lifeboat ethics" is a policy suggested by biologist Garrett Hardin in a world of desperately poor and overcrowded countries.[11] Hardin used the metaphor of lifeboats at sea, some of which are threatened to be swamped by people in the water trying to get in. According to Hardin, the people in the lifeboats that are not completely filled, have three choices. The first is to take in everyone who wants to get on board; but that would lead to the lifeboats being swamped and everyone drowning. The second choice is to take on only a few to fill the empty seats; but that would lead to the loss of the small margin of safety and make for a very difficult decision as to which few will be selected. The third choice is to take no further people on board and to protect against boarding parties. Hardin saw the rich nations of the world as being in partially filled lifeboats and the poor nations as being in overcrowded boats with people spilling into the water because of their inability to control their population growth. Hardin recommends the third choice for the United States. He admits this is probably unjust, but recommends that those who feel guilty about it can trade places with those in the water. Good-willed but basically misguided efforts by the United States to aid poor countries, such as by giving them food during famines, can lead to more suffering in the long run. The emergency food aid contributes to a larger population eventually and thus a deeper crisis in the future.

Growth

Julian Simon, professor of business administration at the University of Maryland, is one of the main spokespersons for the second position on the future, that economic growth will and should continue indefinitely into the future. The rest of this section on "Growth" is my summary of the argument Simon presented in his book *The Ultimate Resource*.[12]

Natural resources are not finite in any real economic sense. When there is a temporary scarcity of a mineral, prices rise and the increased price stimulates new efforts to find more ore and more efficient methods to process it. The higher price also leads to the search for substitutes that are able to provide the same service as the temporarily scarce mineral. In fact, the cost of most minerals has actually been decreasing, so in a real sense minerals are becoming less scarce rather than more scarce. There are large deposits of minerals in the sea and even on the moon that have not yet been tapped. There is no need to conserve natural resources because of the needs of future generations or of poor nations. The present consumption of natural resources stimulates the production of them and improves the efficiency with which they are produced. Both of these developments will aid future generations. Poor nations are not helped by the rich nations using fewer resources; what the poor nations need is economic growth and that growth depends on their increased use of resources.

As with natural resources, the long-run future of energy looks very promising. Aside from temporary price increases caused by the political maneuvering of some countries, the long-run trend of the cost of energy has been downward. Over time, an hour's work has bought more rather than less electricity. This means that energy has become less scarce rather than more scarce. It is likely that an expanding population will speed the development of cheap energy supplies that are almost inexhaustible. In the past, increased demand for energy led to the discovery of new sources, new types of energy, and improved extraction processes.

There is no reason why this trend should not continue into the future. Much of the world has not even been systematically explored for oil.

It is true that the more developed an economy becomes the more pollution it produces, but overall we live in a healthier environment than ever before. The best indicator of the level of pollution is length of life, indicated by the average life expectancy of the population. Life expectancy is rising, not falling, around the world. In the United States and other developed nations it has been rising for the past several centuries, and in the less developed nations for the past several decades. Although a rising income in a country often means more pollution, it also means a greater desire to clean up the pollution and an increased capacity to pay for cleaning it up. If one doesn't believe this, one can compare the cleanliness of streets in rich countries with those in poor countries.

Since World War II, the per capita supply of food in the world has been improving. Famines have become fewer during the past century. The price of wheat has fallen over the long run. The trend toward cheaper grains should continue into the future. Overall, nutrition has been improving and there is no reason why it should not continue to improve into the indefinite future. The amount of agricultural land on the planet has been increasing, especially as irrigation spreads. The amount of arable land is likely to continue to increase and it is not unrealistic to think about land becoming available on other planets. The colonizing of space is not an impossible dream any more.

Additional children mean costs to the society in the short run, but in the long run these children become producers, producing much more than they consume. For both less developed and more developed countries, a moderately growing population is likely to lead to a higher standard of living in the future than is a stationary, or rapidly increasing population. When additional children are born, both the mothers and the fathers work harder and spend less time in leisure. Also, a larger population means a bigger market that makes economies of scale possible; that is, industries can adopt more efficient procedures since they are producing more products for more buyers.

Past studies of animal behavior have often been cited as evidence that crowding is unhealthy, both psychologically and socially, for human beings. This is probably true for animals, but not for human beings. Isolation is what harms human beings, not crowding. In fact, a dense population makes necessary and economical an efficient transportation system. Such a system is essential for economic growth. A dense population also improves communications, something anyone can see by comparing a newspaper in a large city with one produced in a small city. A growing population spurs the adoption of existing technology and the search for new technology as well as the search for and production of new natural resources and energy.

The main question we should ask ourselves when considering population is, What value do we place on human life? Who is to say that the life of a poor person is not of value? Who is to say that a country of 50 million people with a yearly per capita income of $4000 is better than a country with 100 million people and a per capita income of $3000? The most important resource we have on earth is the human mind. The human mind is the source of knowledge. The more human minds we have, the more knowledge we will have to solve the problems we face.

Sustainable Development

A widely accepted definition of sustainable development was given by the World Commission on Environment and Development in its 1987 report, *Our Common Future*: sustainable development enables current generations to "meet their needs without compromising the ability of future generations to meet their own needs."[13]

The term "sustainable development" probably originated with Lester Brown, head of the Worldwatch Institute, a research group established to analyze global problems. In his book, *Building a Sustainable Society*,[14] which was published in 1981 – the same year Julian Simon's book was published – Brown claimed that present economic growth in the world is undermining the carrying capacity of the earth to support life. The Worldwatch Institute believes that if the world does not achieve sustainable development in about 35 years, environmental deterioration and economic decline will probably be feeding on each other, leading to a downward spiral in the human condition. The rest of this section is a summary of what sustainable development could look like in 2030, according to the Worldwatch Institute.[15]

Today, in 2030, birth rates around the world have fallen dramatically, so that the world's population is now about 8 billion instead of 9 billion, which the UN in the mid-1990s had projected for this year. The world's population is either stable or slowly declining to a number that the earth can sustain indefinitely.

Renewable and clean solar energy and geothermal energy have replaced fossil fuels as the main energy sources for the world. The type of solar energy used varies according to the climate and natural resources of the region. Northern Europe is relying heavily on wind power and hydropower. Northern Africa and the Middle East are using direct sunlight as their main energy source. Japan, Indonesia, Iceland, and the Philippines are tapping their ample geothermal energy reserves, while Southeast Asia is using a combination of wood, agricultural wastes, and direct sunlight. Solar thermal plants stretch across the deserts in the United States, North Africa, and central Asia. Most Third World villages now receive electricity from photovoltaic solar cells, which are much less expensive than they were in the mid-1990s. Wind power is common not only in northern Europe but also in central Europe, and vast wind farms exist in the Great Plains of the United States.

The efficient use of energy is widespread. Many of the technologies to achieve high energy efficiency already existed in the mid-1990s. The fuel efficiency of cars has been doubled from that in the mid-1990s, the energy efficiency of lighting systems has been tripled, and typical heating requirements have been cut by 75 percent. Refrigerators, air conditioners, water heaters, and clothes dryers are all highly energy efficient. Improvements in the design of electric motors have made them highly efficient and easy to maintain. Cogeneration, the combined production of heat and power in the same plant, is widespread, thus greatly reducing the use of energy by industry.

The transportation of people has undergone major changes from what it was 35 years ago. The typical European and Japanese cities in the mid-1990s had already developed the first stage of the new urban transport system. Fast and efficient rail and bus systems are used to move the urban population rather than the automobile. When automobiles are used in the cities, they are nonpolluting and highly efficient electric or hydrogen-powered "city cars." Larger cars are available for rent by families for use on long trips. As in some European cities and in many Asian cities in the mid-1990s, bicycles are widely used, for example, by commuters to reach rail lines that are connected to the city center.

People live closer to their work in 2030 than they did 35 years ago, thus contributing to the reduced use of energy. The widespread use of computers to shop from one's home has greatly reduced the use of energy. Telecommunications is permitting many people to work from their homes or in satellite offices, thus greatly reducing the need for travel. Employees and supervisors are connected by computers and other telecommunication devices instead of by crowded highways.

The throwaway mentality that was widespread in the mid-1990s has been replaced by the recycling mentality. Many countries have a comprehensive system of recycling metal, glass, paper, and other materials, with the consumers separating the material for easy collection. The principal source of materials for industry is recycled goods. Waste reduction is greatly aided by pervasive recycling, but even more by the elimination of waste by industry in the production stage. Waste produced by industry has been cut by at least 30 percent from that produced 35 years ago, as many industries have redesigned their industrial processes. Food packaging has been simplified, thus also reducing much waste. And human wastes, after health procedures have been followed to remove the threat of disease, are used in fish farms and in greenbelts surrounding cities for vegetable growing. This practice was already being followed in the mid-1990s in a number of Asian countries. Many households have also helped reduce wastes by composting their yard wastes.

Because the world's population is larger in 2030 than it was in the mid-1990s, the land is used more intensely than it was in the earlier period. Land that was degraded by overuse and neglect is being restored so that it can be brought into use for agriculture. Improved varieties of crops and planting methods are helping conserve fertile soil and water.

215

Major efforts are being taken to halt desertification: one of the best methods has been the eliminating of overgrazing by cutting back on the unnecessarily large herds of animals. Because of the warmer climate produced by greenhouse warming, a larger variety of crops is being grown, many of which are specifically salt-tolerant and drought-resistant.

The forest cover is stable or expanding in most parts of the earth. The cutting down of tropical rainforests has been largely halted and many "extractive reserves" have been established where local people are harvesting nuts, fruits, rubber, and medicines on a sustainable basis. A widely dispersed network of preserves has been established to protect unique and rare animals and plants. The replanting of trees in large areas deforested back in the twentieth century has begun, and efforts are in progress to learn how to harvest the forest for timber without decreasing its productivity, its diversity of species, or its overall health.

Restoring the land has been aided by establishing a pattern of land ownership much more equitable than it was in the mid-1990s. The large land holdings of the few in countries with a majority of poor landless people have been broken up, so that more people now have a chance to earn a livelihood. Also, many government-owned parcels of forests and pastures have been turned over to villages for their management on a sustainable basis.

A shift of employment has taken place so that losses of jobs in coal mining, auto production, road construction, and metals prospecting have been offset by gains in the manufacturing and sale of solar cells, wind turbines, bicycles, mass transport equipment, and recycling technologies.

The trend toward ever larger cities has been reversed. Decentralized, low cost, renewable energy sources have fostered greater local self-reliance and made smaller human settlements more attractive. GNP – the amount of goods and services produced – is no longer accepted as the measure of human progress because it is now seen to undervalue what is needed for a sustainable society. A sustainable society needs qualities such as the durability of products and resource preservation; it does not need waste and planned obsolescence.

As the recognition has grown that environmental threats to peoples' security are greater than military threats, the amount of money spent on the military in the world – which totaled about $ 1 trillion a year in the mid-1990s – has been drastically reduced. This has freed needed funds for energy efficiency, soil conservation, tree planting, family planning, and other sustainable development activities. More countries are now relying on the UN for protection against aggression and have supported the greatly strengthened peacekeeping capabilities of the UN . Some countries have even followed the example of Costa Rica and have abolished their army. At the same time that nations have moved to decentralize power and decision making within their borders, they have also expanded their efforts to cooperate internationally in order to monitor the environment and coordinate efforts to attack global problems.

A slow change in people's values is taking place, so that concern for future generations now has a higher priority than it did earlier. Materialism and consumerism are gradually being replaced by voluntary simplicity, a recognition that personal self-worth cannot be measured by the amount of goods one owns. More efforts are being made to form richer human relationships, and stronger communities, and more emphasis is now placed on music and the arts. As the desire to amass more personal and national wealth has subsided, the gap between the rich and the poor in the world has gradually been reduced. This has led to a reduction in social tensions. There is an expanding recognition in the world of the value of democracy, human rights, diversity, and the freedom to innovate, as these are seen as helping nations achieve sustainability.

Conclusions

Could the world's future contain parts of all three alternative futures presented in this chapter? I believe it could. The dangers that the "doom" alternative conveys are real. Some of them have already taken place in parts of the earth, such as overpopulation in India and China, famine in Africa, and toxic poisoning in the United States, Europe, and Japan. The threat of nuclear weapons is widely recognized. If actions are not taken to reduce the nuclear threat, to reduce population growth in the poorer nations, and to end some forms of environmental deterioration, it is possible that huge loss of life could occur in the future. The positive feature of the doom scenario is that it causes us to recognize real dangers and to try to prevent them from occurring. The negative aspect of this imagined future is that for some people it can weaken their will to act. It frightens them so much that they literally give up on the future, becoming either depressed, numb, or inclined to live for the present. Too much preaching of doom can be a self-fulfilling prophecy if its effect is to discourage action.

Some aspects of the "growth" future appear to be good ones for the developing nations to strive for. They need more economic growth in order to raise their living standards. To attempt to achieve a more equal distribution of income in many of these countries today without further growth would create tremendous political turmoil and would probably bring little benefit to their societies, since in many of these countries there is not that much economic wealth to redistribute. To argue that these countries should not grow economically would be to condemn them to living forever with their present poverty. But to advocate more economic growth for the less developed nations does not deny the need for many of these nations to achieve a more equal distribution of income. Nor does it deny their vital need to slow down their population growth and to move toward a stable population. The less developed nations need to learn from the mistakes of the industrial countries and to make their growth as nonpolluting as possible. The industrial countries have demonstrated well that it is much less costly to a society in the long run to make efforts to prevent environmental destruction from taking place than to try to clean up the damage after it has occurred.

But does more economic growth make sense in the developed nations? It is unpopular today to suggest that it does not, but this may indeed be the case. The

217

desire to acquire more and more material possessions in the wealthy nations has placed a tremendous strain on the planet. This book has been concerned with documenting that strain. The developed nations have achieved for most of their citizens the goal that only kings and queens could achieve in the past – material comfort, an abundance of food, a relatively long life, and leisure. But these countries are finding it very difficult to learn when enough is enough. Obsession with materialism has been condemned by many of the great religions of the world, but it is still an obsession experienced by many. It is clear that this obsession is not healthy – for the individual or for the planet – and that human beings need a different goal for their lives.

The sustainable development future appears to be the one the developed nations should strive for, since the economic growth they are pursuing is destroying the resource base and environment upon which life rests. In the language of economics, it is using up the earth's natural capital – the clean air, clean water, fertile soil, etc. – for a short-term profit. And a basic principle of economics is that if you expend your capital, the financial and physical resources that allow you to produce goods and services, you will soon go bankrupt. It is also clear that this type of development shows little or no concern for future generations. One of the basic rules the Native Americans of the Iroquois confederacy followed in North America before their demise was the rule of the seventh generation: consider how your decisions will affect the lives of the seventh generation to come. Sustainable development is a powerful concept because it is hard to argue against it. How can one publicly defend unconcern for future generations? Also it is a useful concept because it sets up a standard against which one's present actions can be judged.

A sustainable world would not mean the absence of growth, but the growth that would be emphasized would be intellectual, moral, and spiritual growth rather than the growth of material objects. The human race does seem to be at a critical juncture. Will it realize the destructive things it is doing to life on earth and pursue a new course before it is too late? Is it a species that is developing intellectually, morally, and spiritually? The uncertainty of those answers is what makes the present day an exciting and challenging day in which to be alive. The stakes are high. In his novel, *Women in Love*, the British author D. H. Lawrence saw the stakes as follows:

"God cannot do without man." It was a saying of some great French religious teacher. But surely this is false. God can do without man. God could do without the ichthyosaur and the mastodon. These monsters failed creatively to develop, so God, the creative mystery, dispensed with them. In the same way the mystery could dispense with man, should he too fail creatively to change and develop.[16]

For the first time in human history, human beings have the technology to enable them to monitor the planet, to see how their actions are changing the forests, the air, and the water. And they are learning to think of the earth as a single system, a system in which they are just one of the parts. British scientist James Lovelock's Gaia hypothesis, which holds that the earth appears to behave like a single living organism, is an example of this new thinking.[17] Because human beings are the only species to have high intellect and to have knowledge that their time on earth is limited by death, they are learning that they have a special responsibility to all of life on the planet. Whether they are learning this fast enough to prevent irreversible destruction is uncertain.

Although there is little prospect at present of the developed nations adopting sustainable development as their real goal – the one they are putting most of their efforts into achieving – it may come. It became the goal that most nations, at least publicly, subscribed to at the Earth Summit in Rio de Janeiro in 1992. The United States is already moving toward an economy in which occupations that utilize new knowledge will soon be more common than blue-collar manufacturing jobs. And nearly all industrialized nations have already achieved a birth rate that is at or below replacement level, thus leading to a relatively stable population for them in the not-too-distant future.

And there is a slowly growing awareness in the developed nations – as well as in parts of the Third World – that human beings need to live in harmony with nature, to move beyond their compulsion to dominate it. For people who do not yet have this awareness, one of the best ways they can learn this is by personal experience. When they learn, for example, that the water they have been drinking contains cancer-producing chemicals, they learn a lesson about ecology better than any textbook could teach them. Human beings can also learn by using their reason; their use of this capacity can make personal experience less needed as a teaching tool. But either way, human beings can and do learn and change their ways when their own survival depends on it.

We do not know if life exists anywhere else in the universe. It may, on some planets around the billions of stars in our galaxy or in the billions of other galaxies. But even if our present efforts to monitor radio signals from space do reveal other life[18], life as it has developed on earth is probably unique. And preserving this life on our planet in its wonderful diversity and beauty and improving the human condition must be the goal of development. No other will do.

Notes

1   Wendell Berry, "Out of your car, off your horse," *Atlantic Monthly*, 267 (February 1991), p. 63.
2   Jonathan Schell, *The Fate of the Earth* (New York: Avon Books, 1982).
3   Rachel Carson, *Silent Spring* (Greenwich, Conn.: Fawcett Books, 1962).
4   Paul R. Ehrlich, *The Population Bomb*, rev. edn (New York: Ballantine Books, 1971), pp. xi–xii.
5   Paul R. Ehrlich and Anne H. Ehrlich, *The Population Explosion* (New York: Simon and Schuster, 1990).
6   Donella Meadows et al., *The Limits to Growth*, 2nd edn (New York: Universe Books, 1974), p. 24.
7   Donella H. Meadows, Dennis L. Meadows, and Jorgen Randers, *Beyond the Limits: Confronting Global Collapse, Envisioning a Sustainable Future* (Post Mills, Vt.: Chelsea Green Publishing, 1992).
8   Council on Environmental Quality and the Department of State, *The*

Global 2000 Report to the President,: Entering the Twenty-First Century vol. 1 (New York: Penguin Books, 1982), p. 1.

9 Paul Kennedy, *Preparing for the Twenty-First Century* (New York: Random House, 1993).

10 William and Paul Paddock, *Famine-1975!* (Boston: Little, Brown, 1967), 9.

11 Garrett Hardin, "Living on a Lifeboat," *Bioscience*, 24 (October 1974), pp. 561–8.

12 Julian Simon, *The Ultimate Resource* (Princeton: Princeton University Press, 1981).

13 World Commission on Environment and Development, *Our Common Future* (New York: Oxford University Press, 1987).

14 Lester R. Brown, *Building a Sustainable Society* (New York: W. W. Norton, 1981).

15 Lester R. Brown, Christopher Flavin, and Sandra Postel, "Picturing a Sustainable Society," In *State of the World – 1990* (New York: W. W. Norton, 1990), pp. 173 – 90.

16 D. H. Lawrence, *Women in Love* (New York: Viking Press, 1960), p. 470.

17 A description of how the Gaia hypothesis has influenced scientific thinking and how it is evolving is contained in William Stevens, "Evolving Theory Views Earth as a Living Organism," *New York Times*, national edn (August 29, 1989), pp. 17, 19.

18 John Wilford, "Astronomers Open New Search for Alien Life," *New York Times*, national edn (October 6, 1992), pp. B5 & B9.

Further Readings

Adamson, David, *Defending the World* (London: I. B. Tauris, 1990). Adamson expresses doubts about the possibilities for sustainable development and argues that the developed world will be too pressed to save itself to worry about the developing world.

Clark, William C., "Managing Planet Earth," pp. 47–54; Jim MacNeill, "Strategies for Sustainable Economic Development," pp. 155–65; and William D. Ruckelshaus, "Toward a Sustainable World," pp. 166–75, *Scientific American*, 261 (September 1989). In this special issue devoted to exploring the prospects for sustainable human development, Clark focuses on these questions: What kind of planet do we want? and What kind of planet can we get? MacNeill examines how economic growth can be reconciled with the integrity of the environment, and Ruckelshaus describes the policies that can lead to a change in behavior by individuals, industry, and government so they support sustainable development.

Daly, Herman E., and John B. Cobb Jr, *For the Common Good: Redirecting the Economy Toward Community, the Environment, and a Sus-*

tainable Future (Boston: Beacon Press, 1989). An economist and a theologian rarely write a book together, but they do here. After showing how the present growth-oriented Western industrial economies have led to vast environmental destruction, the authors argue for a new economic model that would give moral priority to national and regional communities and to the rights of future generations and nonhuman life.

Durning, Alan, *How Much Is Enough?* (New York: W. W. Norton, 1992). By examining the roots of the consumer society and its negative aspects, Durning argues that the developed countries can and must learn to do with fewer material goods.

Gordon, Anita and David Suzuki, *It's A Matter of Survival* (Cambridge: Harvard University Press, 1991). A comprehensive look at environmental damage and human attitudes toward the earth backs this doomsayer's call for action and a change in our values. The book presents practical suggestions for what individuals and governments can do to create what the authors call a "conserver society."

Kennedy, Paul, *Preparing for the Twenty-First Century* (New York: Random House, 1993). This author of the best-selling *Rise and Fall of the Great Powers*, discusses the major challenges (in demography, technology, and ecology) which he sees the world having to face over the next four to five decades.

Repetto, Robert (ed.), *The Global Possible: Resources, Development, and the New Century* (New Haven: Yale University Press, 1985). Essays written by experts in various environment-related fields discuss the deterioration of resources and the environment and propose practical and political solutions.

Seielstad, George A., *At the Heart of the Web: The Inevitable Genesis of Intelligent Life* (Boston: Harcourt Brace Jovanovich, 1989). Seielstad extends to the whole universe Dr James Lovelock's Gaia hypothesis as he presents the idea that the universe is a living system. By arguing that humans are poised to control the evolution of life on earth, he sees human beings as the shepherds of the earth, responsible for its care.

Timberlake, Lloyd, *Only One Earth: Living for the Future* (London: BBC Books/ Earthscan, 1987). This book offers interesting cases of individuals around the world (in Sri Lanka, Kenya, Solomon Islands, Peru, China, Britain, United States, and Zimbabwe) who are taking action to improve their lives and the environment, and one case (Panama) in which individuals are destroying the environment while they seek economic gain.

World Commission on Environment and Development, *Our Common Future* (New York: Oxford University Press, 1987). After three years of public hearings around the world and testimony by many development experts, this UN appointed commission (commonly called the Brundtland Commission, after its chair) called for the revival of rapid economic growth in the world but in a sustainable fashion and more equitably distributed.

APPENDIX 1

Studying and Teaching Global Issues

For the Student •

You may find it useful to learn how the concept "development" can be used to study global issues and to have an overview of the topics covered in this textbook. One way to do this is to examine the structure of the course I teach in which *Global Issues: An Introduction* is the principal textbook.

Introduction
The first two or three days of the course are spent explaining what "development" means and how development and global issues are related. I define development as economic growth plus the social changes caused by or accompanying that economic growth. In this short introduction I try to help my students understand some of the main differences in the social and economic conditions of the rich and poor countries.

The Wealth and Poverty of Nations
The first full week of the course is spent on getting students to consider the extremely difficult question, Why are some nations rich and some poor? Students examine three of the most widely accepted approaches or views of economic development: the market approach (also called the neoclassical or capitalist approach), the state approach (also called the command economy or socialist approach), and the civil society approach (decentralized development by families, community organizations, and grass roots movements).

223

Population

For two weeks we look at the relationship between population and development. The changing population of the world is described, and the causes of the population explosion in the Third World are given. Students learn how population growth affects development (rapid population growth hinders development by putting a large stress on resources, health and education facilities, the environment, etc.) and how development affects population growth (development at first makes it greater as it lowers death rates but later it reduces birth rates as the education level of women increases and children become less desirable economically and socially). The demographic transition is explained and students become familiar with the factors that lower birth rates. Some attention during this period is paid to the population policies of major countries, such as China. This segment of the course ends with a consideration of the future – whether a stabilization of the world's population will occur and whether the carrying capacity of the earth will be exceeded.

Food

For two weeks food holds our attention. World food production trends are examined, and a tentative answer is given to the question of how many are hungry in the world today. We investigate the causes of hunger in parts of the Third World (poverty – the lack of development – is one cause). Students learn how the availability and quality of food affect development (malnourished people are not good producers) and how development affects both the production of food and the type of food consumed (industrialized agriculture produces a large amount of food, but wealthy people often do not have a healthy diet). A short history of the Green Revolution is given. The food policies of the United States and a few other countries are examined. Finally, if time permits, we think about how changes in the climate, the amount of arable land, and the cost of energy could affect future food supplies.

Energy

Two weeks is not enough time to investigate thoroughly the relationship between energy and development, but it is enough time to introduce students to this vital subject. A description of the energy crisis caused by the developed world's dependency on a polluting and highly insecure energy source – oil – is followed by a summary of the responses to that crisis by the United States, Western Europe, and Japan. The effect of the energy crisis on Third World development plans is explained. As we explore the relationship between energy use and development, students learn about the shift in the types of energy sources that took place as the Industrial Revolution progressed and how there has been a partial decoupling of energy consumption and economic growth – a new ability to produce economic growth with less energy. The subject of the greenhouse effect (global warming) could be examined in the next section of the course, but I include it in the energy section because it serves as a good bridge to my discussion of nonrenewable and renewable energy sources. The role of conservation during the present period of energy transition is also explored. I end this section of the course with a presentation of the main arguments for and against nuclear power, which allows me to demonstrate how difficult and complicated are the choices the political system must make when dealing with energy.

The Environment

As is the case with energy, two weeks is not much time to explore the effects development has on the environment – and the reverse, the effects the environment has on development – but significant information on the subject can be passed to students. (Poor people are hard on the environment as they struggle to survive, but the rich may or may not treat the environment well.) A brief history of the awakening in the United States to threats to the environment caused by industrialization introduces the subject, and provides the setting for an examination of the threats to the air, water, and land that have come with development. Airborne lead, acid rain, and the depletion of the ozone layer illustrate some of the main concerns we have at present with air pollution. The current concern with threats to our groundwater by migrating chemicals presents an example of water pollution caused by development, and the problem of how to handle huge amounts of solid and toxic wastes demonstrates well to the students the extremely difficult tasks the political system faces as it tries to preserve the land.

The problem of deforestation in the Third World is briefly examined so that students become aware of the harm deforestation can bring to the land, its connection to the extinction of species, as well as the changes it can make in the climate. The connection between development and the extinction of cultures is also examined as, for example, in the name of development the forest homes of numerous indigenous peoples are being destroyed. Chemicals, cancer, and pesticides are considered under a section in which we focus on the workplace and the home. Finally, the effect development has on the use of natural resources gains our attention, and students learn – often with some surprise – that development has often, at least so far, made many natural resources more available and cheaper.

Recycling, substitution, and the mining of low grade ores are subjects presented at the end of this section of the course as we consider future supplies. The concept "overdevelopment" (consuming and polluting at a rate that cannot be maintained indefinitely) is also presented, as students consider reducing needs as a possible response to scarcities.

Technology

To many people, technology and development are synonymous. Technology is what makes economic growth and social change happen. Students are reminded of the many benefits that technology has brought to

225

our lives. But because they are more aware of the benefits than the harm technology can produce, the course focuses on the dangers. Students learn that the decision of whether or not to use a certain technology can be a difficult one, especially in "tragedy of the commons" situations where short-term interests and long-term interests conflict. Illustrations of the unanticipated consequences of the use of technology are given as are examples of the inappropriate uses of technology. Limits to the "technological fix" are illustrated. The issue of war is introduced with technology making the destructive capacity of weapons greater. The threat of nuclear weapons is presented as a case study under the technology section.

Alternative Futures

I end the course by focusing on different possible futures. During the last week we examine the main arguments that advocates make for the possibilities that our present type of development is leading us to "doom," or to continued "growth," or to "sustainable development" in the future.

For the Teacher

The Problem

Improving and increasing international studies has become a priority on many campuses,[1] but as a report for the American Council on Education concludes, "the internationalizing of undergraduate education still has a long way to go."[2] How far it has to go can be easily shown. Reports of the shocking ignorance of people in the United States about other countries are well known, but less well known, and of some embarrassment to the college teaching profession, is that college-age people in the country are the most ignorant of all adults. Adding to the insult is the fact that attending four years of college reduces that ignorance only slightly.[3] Eighteen to twenty-four-year-olds in the United States in the late 1980s possessed *less* information about the world than the same age group had 40 years earlier.[4]

This information is especially surprising given the new emphasis many colleges are placing on international studies. Also surprising is the fact that the average student in a four-year college or university takes several international studies courses, outside of foreign language instruction, before he or she graduates.[5] But a close look at these international studies courses reveals that most of them still focus on only one country or one region (often Western Europe), and only a few focus on a problem or issue that is found throughout the world. Also, few are interdisciplinary, and *only a minority deal with the world as it is today*.[6]

A national study of international courses at both the precollegiate and collegiate level in the mid-1980s found that most textbook publishers were producing books that reportedly dealt with global issues, but in reality many of these textbooks were retitled versions of older texts.[7] That report concluded that because textbooks change very slowly, there is a "need for model or exemplary materials that are sensitive to the realities of a global society."[8]

We indeed seem to be far from achieving what one report called an important characteristic of the truly internationalized university: it is a school where "no student graduates who has never been asked to think about the rights and responsibilities of this country in the world community, or who has never been brought to empathize with people of a different culture."[9]

Preparing students so that they will be able to function in an increasingly complex and interdependent world is a huge task, one which will require a better trained and more committed faculty and college administration. No easy answers, solutions, or quick fixes are possible, but many different methods and approaches are being tried, with varied degrees of success. As the American Council on Education study found, what we do not have now in the United States is a way to know what works, and what does not, and why it does or does not.[10] Needed are reports of successes and failures in the attempts to achieve the important characteristic of the truly internationalized university that the above quotation appropriately identifies.

A Solution

While attending a conference on the Third World, I heard college teachers complain that they could not get their students interested in studying the Third World, where most of the world's people live. As I thought about this complaint, I realized that I had discovered an answer to the question, How do you get American students to want to study the non-Western world? I know that you don't do it by reminding them that their bananas come from that world. The student's reaction to that statement is, So what? Who cares? The way you get them interested is by introducing real global problems and exploring their possible solutions. You demonstrate that global problems are American problems, that our actions help create or solve the problems, and that the problems affect our lives, in the present as well as in the future.

Over the past 14 years I have taught a course for undergraduates called "Global Issues and Development." The course, outlined in the section addressed to the student, focuses on many of the most important global issues today, issues that both the more developed and the less developed nations can no longer ignore.

I believe that one reason many social science teachers do not teach a course on global issues is that they do not know how to deal with these issues in a respectable, scholarly way – in a manner that will prevent the class from becoming just a forum for the discussion of current events. But I have found that there is a concept – "development" – which can serve as the tool we need for treating these issues in a responsible manner. Social scientists commonly use this concept only with reference to

227

the poorer nations, but "development" can also be a powerful tool for analyzing conditions in and actions of the richer nations.

Teaching Techniques

How does one teach the above material? I have used a combination of techniques. I have adopted as the basic textbook this book *Global Issues: An Introduction*. I have also used the latest edition of the Worldwatch Institute's *State of the World*.[11] This book is an excellent annual updating of many of the topics covered in my course, although the large amount of detailed, factual information it contains overwhelms some undergraduates. At times, in place of *State of the World*, I have used the United Nations Development Programme's *Human Development Report*, which covers many development-related subjects.[12] Students read selections from the latest edition of *Annual Editions: Global Issues*, which is a collection of articles from many different sources – some with opposing viewpoints – on many of the issues presented in the course.[13] Students are also required to subscribe to the *Christian Science Monitor*, which allows them to follow current developments in all of the subjects covered in the course. Weekly quizzes are given on the *Monitor*.

All possible examination questions are given to the students (see Appendix 3) and we use these questions to guide our discussion of the textbook. I do not give lectures. The questions on the examinations are randomly selected from these questions. I find that students learn the material better when they know what they will be tested on.

Videotapes play an important role in the course. Many excellent programs related to topics in our course appear on public television (see Appendix 2). The experience of seeing an interesting, current portrayal of a topic we are studying is a powerful teaching technique. The tapes reinforce what the students are learning and broaden their knowledge. Also, the tapes serve another important role. Studying global issues can be depressing. The problems are numerous and serious, and at first glance appear to be unsolvable. The tapes help counter that depression by often showing what some individuals are doing to attack these problems. I try to show at least two tapes related to each of the five main subtopics in the course.

Students write a five- to eight-page typewritten research paper. In the paper they focus on an issue in greater depth than we have been able to in the course.

A course of instruction following the above outline utilizes three levels of analysis, which contribute to its effectiveness: the individual, the nation, and the international system. To understand the issues one must look at the behavior of individuals, the actions and policies of nations, and the condition of the world's environment as well as of its economic and political systems. Solutions to the global problems require individual efforts, new national policies, and international agreements.

Such a course of instruction has three main goals. The first is to increase student knowledge of some of the most important problems facing the world today, a knowledge that the student learns comes from many different disciplines. The second goal is to help students learn of the complex interrelationships among the issues. The third is to evaluate possible solutions to the problems studied. As the students consider possible solutions, they learn the vital fact that human actions (including their own) can change the world in very different ways.

Can these goals be achieved? Certainly they cannot for every student, nor will every student who achieves one, achieve all three. But many can achieve one or more of these goals. Students appreciate an effort that helps them understand the complicated and rapidly changing world in which they live. When we help them acquire this information, we are giving them both the knowledge they will need to live in today's world, but more importantly, the knowledge that will enable them, if they so desire, to add their talents to the efforts being made to solve many of these global problems.

Student Comments

For the past 12 years I have at the end of the course asked students to write, in a short unsigned essay, what they felt was the most important thing they learned in the course. These three responses give some common conclusions:

> The most important thing I learned is that problems concerning population, food, energy, etc. are <u>real</u>. I feel that most people don't realize the magnitude of these problems. However, by taking this course, I now see that all these problems are greater than I originally thought … This course taught me the first step in combating these problems, and that is to recognize that they are REAL!

> I had… known about the environmental movement and even considered myself an environmentalist. Sure I wanted to take care of my environment; new energy sources sounded cool; pollution was bad and needed to be stopped, etc. However, I never really knew how <u>interconnected</u> all of this was until I took this course… . I learned how changes in one area can drastically affect what I previously thought were unrelated things… . I learned that all of these problems are interconnected and must be studied as such if any real (long term) solution is ever to be found for them.

> The most important thing I learned was to stop thinking like an American and only think about self-interest. Rather now I think about my neighbor be it in Converse Heights or my neighbor in South America. Professor Seitz, you focused my mind to look at the big picture instead of the small one. When I… [threw away an empty] can of Coke previously I would say, 'What can I do about recycling?' Now I see that even a little effort to make a difference does just that, it makes a difference. Now when I get in my car to go to the store, I

think twice and now I usually will walk. Before when I said [what's wrong] with one more light on, it's just 20 cents a day lost. Now I think about how [the production of] electricity pollutes the atmosphere, so now I conserve electricity and other fossil fuels as well. To sum it all up, I have learned to be more responsible to this precious world we call earth. For that, whatever grade I receive, I thank you for opening not just my eyes but my mind.

Notes

1 Ann Kelleher, "One World, Many Voices," *Liberal Education* 77 (November/December 1991), pp. 2–7.

2 Richard D. Lambert, *International Studies and the Undergraduate* (Washington: American Council on Education, 1989), p. 153.

3 Ibid., p. 107.

4 Ibid., p. 106.

5 Ibid., p. 126.

6 Ibid., pp. 115–27.

7 Andrew F. Smith and Walter Brown, *Research on Learning Packages and Course Syllabi Developed for International Studies Courses: A Collection, Analysis and Dissemination Project* (1986) (ERIC, a US Department of Education data base, Document ED 288746), p. 15.

8 Ibid.

9 Humphrey Tonkin and Jane Edwards, "Internationalizing the University: The Arduous Road to Euphoria," *Educational Record* 71 (Spring 1990), p. 15.

10 Lambert, *International Studies*, p. 157.

11 Lester Brown et al., *State of the World* (New York: W. W. Norton, annual).

12 United Nations Development Programme, *World Development Report* (New York: Oxford University Press, annual).

13 Robert M. Jackson (ed.) *Annual Editions: Global Issues* (Guilford, Conn.: Dushkin Publishing, annual).

APPENDIX 2

Suggested Videos

After the Warming [global warming], Les Eaton, executive producer. Maryland Public Television, 1990.

Amazonia: The Road to the End of the Forest [the burning of the rain forest, failed settlements, and Chico Mendes interview], produced by Canadian Broadcasting Corporation, 1990.

Arctic to Amazonia [indigenous people discuss environmental threats to their lands], Erik van Lennep, producer. Turning Tide Productions, 1993.

Back to Chernobyl, *Nova*, Paula S. Apsell, executive producer. Produced by Kurtis productions and WGBH-Boston for Nova, 1989.

Black Triangle: Eastern Europe [major polluted area where Poland, Czech Republic, Slovakia, and Germany meet], A Central Television Production, 1991.

Chelyabinsk: The Most Contaminated Spot on the Planet [Russian nuclear weapons production site], Slawomir Grunberg , director, 1994.

Children of Chernobyl Produced by Yorkshire Television, 1992.

China's Only Child [population control], *Nova*, John Mansfield, executive producer, WGBH-Boston/BBC, 1984.

Columbus Didn't Discover Us [cultural survival of American indigenous peoples], Robbie Leppzer, director. Turning Tide Productions, 1992.

Common Ground [organic farming], Christopher N. Palmer, executive producer. Chedd-Angier Productions, 1987.

Contact: The Yanomami Indians of Brazil. Produced by Geoffrey O'Connor, Realis Pictures, Inc., 1990.

The Cosmic Joke, in two parts [population explosion: focuses on Indonesia, Mexico, Africa], Dr June Goodfield, producer. Ambrose Video Publishers, 1990.

The Endangered Earth: The Politics of Acid Rain. Films for the Humanities and Sciences video distributors, 1991.

The Energy Alternative series: A Global Perspective, in three parts [innovative solutions to the energy crisis] Produced by Grampian Television and INcA, 1990.
  1 Changing the Way the World Works
  2 The Rich Get Richer
  3 Power to the People

Environment at Issue [presents the debate over how damaging environmental conditions are likely to become and what should be done about them] Produced and directed by Jeffrey Tuchman for the Public Agenda Foundation, 1991.

The Environmental Tourist, Christopher N. Palmer, executive producer. Produced by the National Audubon Society, Turner Broadcasting, and WETA-Washington, DC, 1991.

From the Heart of the World [pollution and the extinction of cultures], *Nature*, David Heeley, Fred Kaufman, executive producers. For BBC, in association with the Goldman Foundation, 1991.

Future Talk [how actions today will affect the future, with Bill Moyers]. Films for the Humanities and Sciences video distributors, 1994.

Gertrude Blom: Guardian of the Rain Forest [the story of the life of a Swiss-born conservationist]. Produced by Cinta Productions, 1989.

Global Dumping Ground [toxic waste dumping in the Third World], *Frontline*. Lowell Bergman, Center for Investigative Reporting, 1990.

Global Links: A Close-up Look at Third World Development, in six parts, Jaime Martin-Escobal, producer/director. WETA-TV and the World Bank, 1988.
  1 Traditions and the 20th Century [impact of economic initiatives on traditional societies]

2 Curse of the Tropics [public health]
3 Women in the Third World
4 Earth – the Changing Environment
5 The Urban Dilemma [rapidly growing cities]
6 Education – A Chance for a Better World

Global Report: Can Tropical Rain Forests be Saved? Richter Productions, 1992.

Goddess of the Earth [the Gaia hypothesis], *Nova*, Paula S. Apsell, executive producer, 1986.

The Green Iguana: A New Hope for the Rain Forest [raising iguanas for profit to protect the rain forest]. Produced by Norwegian Broadcasting Corporation, 1992.

Is It Hot Enough? [greenhouse effect and global warming], *Nova*, Paula S. Apsell, executive producer. Produced by BBC and WGBH-Boston for Nova, 1989.

Halting the Fires [burning of the Amazon rain forest]. Produced by Channel 4, England, 1990.

Hot or Not: The Global Greenhouse Debate, Les Eaton, executive producer. Maryland Public Television, 1991.

Hot Wiring America's Farms [energy use and sustainable agriculture]. Front Range Educational Media Corp. and KBDI-TV Denver, 1988.

Human Tide [explosive growth of world population]. Produced by Canadian Broadcasting Corporation, 1994.

Human Wastes: Turning Waste into a Resource. Produced by John Blake Associates Ltd., for Channel 4, 1992.

Hungry for Profit [agribusiness in the Third World]. Robert Richter, executive producer, for Non-Fiction Television.

The Legacy of Nuclear Bomb Making [cleaning up nuclear weapons production facilities]. Center for Defense Information, 1994.

Local Heroes, Global Change, in four parts [on overcoming poverty and creating change], David Kuhn, executive producer. World Development Productions, 1990.
    1 With Our Own Eyes [development by local people and institutions such as by the Grameen Bank, Bangladesh]
    2 Against the Odds [barriers to development with examples from

233

Jamaica and Ghana]
3   Power to Change [focuses on how people emerge from under-development, with examples from India and Bolivia]
4   The Global Connection [explores some proposed changes in the relations between the North and the South which could encourage development]

Millennium: Tribal Wisdom and the Modern World, in ten parts [what modern societies can learn from tribal societies], Adrian Malone, executive producer. Biniman Productions, KCET-TV (Los Angeles) /BBC, 1992.

1 Shock of the Other	6 Touching the Timeless
2 Strange Relations	7 A Poor Man Shames Us All
3 Mistaken Identity	8 Inventing Reality
4 An Ecology of Mind	9 The Tightrope of Power
5 The Art of Living	10 At the Threshold

Nuclear Power: The Hot Debate. Produced by Canadian Broadcasting Corporation, 1990.

Ozone: The Hole Story, Bill Kurtis, executive producer. Produced by Kurtis Productions, Ltd. and WTTV-Chicago, 1992.

Planet Earth: Fate of the Earth [the last program in a series, focuses on the creation of life; threats caused by nuclear weapons, deforestation, global warming; and hope for the future], Thomas Skinner, executive producer. WQED-Pittsburgh, in association with the National Academy of Sciences, 1986.

Politics of Power, *Frontline*, David Fanning, executive producer. Produced by WGBH-Boston and the Center for Investigative Reporting, 1992.

Politics, Power, and Pollution [relationship between corporate productivity and environmental responsibility, with Bill Moyers]. Films for the Humanities and Sciences video distributors, 1994.

The Politics of Trees [controversy over how to manage old-growth forests of the US Pacific Northwest, with Bill Moyers]. Films for the Humanities and Sciences video distributors, 1994.

Profit the Earth, [free market environmentalism: using the market to achieve environmental goals], Jerry M. Landay and Ron Hall, executive producers. Produced by the Nebraska Educational Television Network, 1990.

The Quiet Revolution {overcoming poverty, inequality, and ecological devastation in Bangladesh, Honduras, India, Nepal, the United States, and Zimbabwe]. Produced by South Carolina Educational Television, 1995.

Race to Save the Planet, in ten parts, John Angier, executive producer. WGBH-Boston Science unit, in association with Chedd-Angier, 1990.
    1 The Environmental Revolution
    2 Only One Atmosphere [ozone depletion and greenhouse effect]
    3 Do We Really Want to Live This Way? [pollution]
    4 In the Name of Progress [Is there an inevitable conflict between economic development and environmental protection?]
    5 Remnants of Eden [extinction of species]
    6 More for Less [energy efficiency]
    7 Save the Earth – Feed the World
    8 Waste Not, Want Not [solid waste and recycling]
    9 It Needs Political Decisions [environmental politics]
    10 Now or Never [environmental action]

Radioactive Waste Disposal: The 10,000-Year Test. Films for the Humanities and Sciences video distributors, 1993.

Rocking the Boat: You Can Fight City Hall [effective coalitions are formed by ordinary citizens to fight local pollution]. Films for the Humanities and Sciences video distributors, 1993.

Sea of Oil [*Exxon Valdez* oil spill], M. R. Katzke, director. Affinity Films, 1990.

Spaceship Earth: The Watch Keepers [the last in a ten-part series calls for the balancing of the needs of the environment with the economic needs of people], Nicholas Barton, executive producer. Network Television/Antelope films Production, for Channel Four Television, in association with South Carolina Educational Television, 1991.

Steering into the Future: New Models of Transportation. Produced by Roynn Lisa Simmons for Connecticut Public Television, 1993.

The UN's Nuclear Detective [International Atomic Energy Agency tries to prevent the misuse of nuclear materials for military purposes]. Center for Defense Information, 1993.

Water Wars, in three parts ["Good as Gold" looks at scarcity of water in the US Southwest; "To the Last Drop" describes tensions in the Middle East created by the need for water; and "The Giver of Life" discusses troubles over water in the former Soviet Union.] Produced by the BBC, 1992.

Video Distributors

The following is a partial list of relevant video distributors:

1 Filmakers Library, 124 East 40th Street, New York, NY 10016; telephone (212)808–4980.

2 Films for the Humanities and Sciences, PO Box 2053, Princeton, NJ 08543-2053; telephone: (800)257–5126.

3 Turning Tide Productions, PO Box 864, Wendell, MA 01379; telephone (800)557–6414.

4 Ambrose Video Publishing, Inc., 1290 Avenue of the Americas, Suite 2245, New York, NY 10104; telephone: (800)526–4663.

5 Video Finders, a service of Public Broadcasting Service station KCET/Los Angeles; telephone (800)328–PBS1. Provides availability and ordering information on 70,000 videocassette titles, including 3,000 programs broadcast by PBS.

Information on how to rent or purchase available videos is contained in a reference book found in many libraries: *Bowker's Complete Video Directory 1993*, which is published by R. R. Bowker, A Reed Reference Publishing Company, New Providence, New Jersey.

Study and Discussion Questions for Students and Teachers

Chapter 1. The Wealth and Poverty of Nations ● ● ● ● ● ● ● ● ● ● ●

1   Discuss the growing gap between the rich and poor on earth and the improvement of living standards for many, giving examples and explaining why these two trends are occurring.
2   Explain what the market approach to economic development says about why some nations are rich and some are poor.
3   Discuss some of the main arguments that advocates for the market approach make and some of the main criticisms of the approach.
4   Explain what the state approach to economic development says about why some nations are rich and some are poor.
5   Discuss some of the main arguments that advocates for the state approach make and some of the main criticisms of the approach.
6   Explain how the civil society approach would promote development in the poorest nations.
7   Discuss some of the main arguments that advocates for the civil society approach make and some of the main criticisms of the approach.
8   What are the main conclusions the author draws about the relative strengths and weaknesses of the market, state, and civil society approaches, and give some of the main reasons he comes to these conclusions.

Chapter 2. Population ● ● ● ● ● ● ● ● ● ● ● ● ● ● ● ● ● ● ●

1   In a discussion of the changing population of the world, explain what is happening, where growth is taking place, and why there is movement of population within Third World nations.
2   Discuss the causes of the population explosion, and explain why

birth rates are high in less developed nations.

3   Discuss some of the main negative features of a rapidly growing population.

4   Discuss the main problems that can be created when a country's population growth is low and when the elderly begin to increase in number.

5   Discuss what UN conferences have concluded about the relationship between population growth and poverty, and how population growth can be reduced.

6   Explain the demographic transition and show the differences between the experiences of the more-developed and the less-developed nations with the transition.

7   Discuss the factors that cause birth rates to decline.

8   Discuss governmental policies that are designed to control the growth of population; include the experiences of Mexico, Japan, India, and China.

9   Explain why some governments have wanted to promote the population growth in their countries, giving some specific examples of such countries and their reasons for having such a policy.

10 When and at what level will the world's population probably stabilize? In a discussion of whether the planet will be able to support this population, explain and use the concept "carrying capacity."

11 What are some of the factors that must be considered when one tries to answer the question "What is the optimum size of the world's population?" Identify some of the most likely population-related problems that may occur in the future.

Chapter 3. Food ● ● ● ● ● ● ● ● ● ● ● ● ● ● ● ● ●

1   Discuss how much food is being produced at present and how many people are hungry. Who are the hungry and where do they live?

2   Discuss the causes of world hunger.

3   Discuss how the availability of food affects development in the Third World.

4   Discuss how development has affected the amount of food produced in the developed nations and how food is produced.

5   Discuss some of the most important negative features of modern Western agriculture.

6   Explain how a nation's diet changes as the nation develops, and some of the more harmful characteristics of this diet.

7   Explain what the Green Revolution is, and identify some of its most important positive and negative features.

8   Explain why many less developed nations have not placed a high priority on rural development and the development of agriculture in their countries.

9   Discuss the governmental policies toward agriculture and rural

development of the following: Northeast Asian countries (Japan, South Korea, Taiwan – treat these three as one), China, the former Soviet Union, United States.

10 Discuss how climate, amount of arable land, energy costs, and increased efficiency could affect future food supplies.

11 Discuss how new technology (such as biotechnology), fishing, and aquaculture could affect future food supplies. Explain why the prospects for food production in the future are both promising and troubling.

Chapter 4. Energy ●

1 In a discussion of the energy crisis, explain what it is, and explain the first, second, and third oil shocks.

2 Discuss the responses by the US government to the energy crisis.

3 Discuss the responses by the governments of Western Europe, Japan, and China to the energy crisis.

4 Explain how the energy crisis has affected Third World development plans.

5 Discuss the shift in sources of energy and the increase in the use of energy that have taken place as development occurred in the world.

6 In a discussion of the relationship between economic growth and energy growth, explain what the past one-to-one relationship was, why it has changed in the developed nations, and why energy efficiency has been less in the United States than in Europe and Japan.

7 Discuss global warming (the greenhouse effect), explaining what it is, what is causing it, what it might do, and what can be done about it.

8 Discuss the main nonrenewable sources of energy, indicating some of their most important positive and negative features.

9 Discuss the main renewable sources of energy, indicating some of their most important positive and negative features.

10 In a discussion of conservation as a form of energy, explain why it can be considered a source of energy, how it works, and its advantages.

11 Briefly explain the main differences between fission, fusion, and fast-breeder reactors, and describe the present status of nuclear power around the world, giving reasons for this situation.

12 Discuss the main arguments for withdrawing governmental support for nuclear power.

13 Discuss the main arguments for continuing governmental support for nuclear power.

Chapter 5. The Environment ● ● ● ● ● ● ● ● ● ● ● ● ●

1 Discuss how the three UN conferences on the environment since 1972 show changing attitudes toward the environment in the world.
2 Discuss how development produces air pollution and give some evidence of the harmful effects of this pollution. Include in your answer the negative effects of airborne lead.
3 Explain acid rain and its causes. What effects does it have, and what actions are being taken to combat it?
4 Identify the problems generated by the depletion of the ozone layer and its causes. How extensive is this depletion? What actions are nations taking to combat it?
5 Discuss how development has affected the water on the planet, including the drinking water in the United States and in some European countries.
6 Discuss the relationship between development and production of wastes, both solid and toxic, and identify some of the ways governments can control wastes.
7 What is causing deforestation? Where is it occurring? How extensive is it, and what harm is it doing?
8 What are the main causes of cancer and how are they related to development? What is the relationship between the production of artificial substances, such as chemicals, and cancer?
9 What problems do pesticides cause? Explain how and where these problems arise.
10 How has development affected the availability and price of nonfuel natural resources? Why has this happened?
11 Discuss three of the four steps a country can take to counteract the shortages of a needed material if it cannot locate new rich deposits of the ore.
12 What is causing the extinction of species? Where is it mostly taking place? What harm does it do, and how can it be combated?
13 In a discussion of the extinction of cultures, explain why it is bad and its relationship to development. What can be done to combat it? Give an example of a culture under siege.
14 Discuss what makes environmental politics so controversial.

Chapter 6. Technology ● ● ● ● ● ● ● ● ● ● ● ● ● ●

1 What is the "tragedy of the commons"? How do some present conflicts between short-term versus long-term benefits from using a technology represent this type of situation?
2 Discuss the unanticipated consequences of the use of technology. Include in your discussion the use of DDT, factory farms, and the Green Revolution.

3 Discuss the inappropriate uses of technology, showing why intermediate or appropriate technology in the Third World could be more
beneficial than high or hard technology. Discuss another situation
that illustrates the inappropriate use of technology.

4 Explain the meaning of a "technological fix" and give two illustrations of its limits.

5 Discuss why war can be considered a problem in the use of technology, giving special attention to the characteristics of modern war.

6 Explain how the threat of nuclear weapons can be considered a
problem in the use of technology.

7 Discuss the new dangers that the world faces in the post–Cold War
period because of nuclear weapons.

Chapter 7. Alternative Futures ● ● ● ● ● ● ● ● ● ● ● ●

1 Discuss the most commonly mentioned disasters included in the
"doom" future.

2 Discuss the policies of triage and lifeboat ethics that some have recommended as ways to deal with doom situations.

3 Discuss the main components of the "growth" future.

4 Discuss the main components of the "sustainable development"
future.

5 Explain why the author believes the world's future may contain parts
of all three alternative futures presented in the final chapter.

Selected Bibliography

Abelson, Philip H., "Uncertainties About Global Warming," *Science*, 247 (March 30, 1990), p. 1529.

Adams, Ruth, and Susan Cullen, (eds), *The Final Epidemic: Physicians and Scientists on Nuclear War* (Chicago: Bulletin of the Atomic Scientists, 1981).

Adelman, Irma, and Cynthia Taft Morris, *Economic Growth and Social Equity in Developing Countries* (Stanford: Stanford University Press, 1973).

Alauddin, Mohammad, and Clement Tisdell, *The "Green Revolution" and Economic Development* (New York: St Martin's Press, 1991).

Albright, David, "Chernobyl and the US Nuclear Industry," *Bulletin of the Atomic Scientists*, 42 (November 1986), pp. 38–40. "Armed Force and Imported Resources," *The Defense Monitor*, 11, no. 2 (1992), pp. 1–4.

Arms, Suzanne, *Immaculate Deception: A New Look at Women and Childbirth in America* (Westport, Conn.: Bergin and Garvey, 1984).

Attenborough, David, *The Living Planet: A Portrait of the Earth* (Boston: Little, Brown, 1984).

Barringer, Felicity, "Chernobyl: Five Years Later the Danger Persists," *New York Times Magazine* (April 14, 1991), pp. 28–74.

Baum, Warren C., and Stokes M. Tolbert, *Investing in Development: Lessons of World Bank Experience* (Oxford: Oxford University Press, 1985).

Bazzaz, Fakhri A., and Eric D. Fajer, "Plant Life in a CO_2-Rich World," *Scientific American*, 266 (January 1992), pp. 68–74.

Berger, Peter, "Underdevelopment Revisited," *Commentary*, 78 (July 1984), pp. 41–5.

Bingham, Sam, "Czechoslovakian Landscapes," *Audubon* (January 1991), pp. 92–103.

Blair, Bruce G., and Henry W. Kendall, "Accidental Nuclear War," *Scientific American*, 263 (December 1990), pp. 53–8.

Botkin, Daniel B., "A New Balance of Nature," *Wilson Quarterly* (Spring 1991), pp. 61–72.

Bretton, Henry L., *International Relations in the Nuclear Age: One World Difficult to Manage* (Albany: State University of New York Press, 1986).

Broecker, Wallace S., "Global Warming on Trial," *Natural History* (April 1992), pp. 6–14.

Brown, Lester R., "A New Era Unfolds." In Lester R. Brown et al., *State of the World – 1993* (New York: W. W. Norton, 1993), pp. 3–21.

—, *Building a Sustainable Society* (New York: W. W. Norton, 1981).

Brown, Lester R., and Edward Wolf, "Food Crisis in Africa," *Natural History*, 93 (June 1984), pp. 16–20.

Brown, Lester R., et al., *State of the World* (New York: W. W. Norton, annual).

Brown, Lester R., Christopher Flavin, and Sandra Postel, In "Picturing a Sustainable Society," *State of the World – 1990* (New York: W. W. Norton, 1990), pp. 173–90.

Brown, Michael, *Laying Waste: The Poisoning of America by Toxic Wastes* (New York: Pantheon Books, 1980).

Caldwell, John C., and Pat Caldwell, "High Fertility in Sub-Saharan Africa," *Scientific American*, 262 (May 1990), pp. 118–25.

Carson, Rachel, *Silent Spring* (Greenwich, Conn.: Fawcett Books, 1962).

Chiles, James R., "Tomorrow's Energy Today," *Audubon* (January 1990), pp.59–72.

Cline, Stephen, "Down on the Fish Farm," *Sierra* 74 (March/April 1989), pp. 30–8.

Conn, Robert W., et al., "The International Thermonuclear Experimental Reactor," *Scientific American* , 266 (April 1992), pp. 103–110.

Conquest, Robert, *The Harvest of Sorrow: Soviet Collectivization and the Terror-Famine* (New York: Oxford University Press, 1986).

Council on Environmental Quality and the Department of State, *The Global 2000 Report to the President: Entering the Twenty-First Century*, vol. 1 (Washington: Government Printing Office, 1980; New York: Penguin Books, 1982).

Critchfield, Richard, *The Villagers: Changed Values, Altered Lives: The Closing of the Urban-Rural Gap* (Garden City, N.Y.: Anchor Press/Doubleday, 1994).

—, "Science and the Villager: The Last Sleeper Wakes," *Foreign Affairs*, 61 (Fall 1982), pp. 14-41.

—, *Villages* (Garden City, NY: Anchor Press/Doubleday, 1981).

Davis-Floyd, Robbie E., *Birth as an American Rite of Passage* (Berkeley: University of California Press, 1992).

Del Tredici, Robert, *At Work in the Fields of the Bomb*, (New York: Perennial Library/Harper and Row, 1987).

Deudney, Daniel, and Christopher Flavin, *Renewable Energy* (New York: W. W. Norton, 1983).

Deutch, John M., "The New Nuclear Threat," [nuclear proliferation] *Foreign Affairs*, 71 (Fall 1992), pp. 120–34.

DeWalt, Billie, "The Cattle Are Eating the Forest," *Bulletin of the Atomic Scientists*, 39 (January 1983), pp. 18–23.

Donaldson, Peter J., and Amy Ong Tsui, "The International Family Planning Movement," *Population Bulletin*, 45 (Washington: Population Reference Bureau, November 1990), pp. 1–47.

Dreze, Jean, and Amartya Sen, *Hunger and Public Action* (Oxford and New York: Clarendon Press, 1989).

Dunlap, Thomas R., *DDT: Scientists, Citizens, and Public Policy* (Princeton: Princeton University Press, 1981).

Durant, Mary, "Here We Go A-Bottling," [recycling] *Audubon*, 88 (May 1986), pp. 32–5.

Durning, Alan B., "Supporting Indigenous Peoples." In Lester Brown et al. *State of the World–1993* (New York: W. W. Norton, 1993), pp. 80–100.

—, "Ending Poverty," *State of the World*, 1990 (New York: W. W. Norton, 1990), pp. 135–53.

—, *How Much Is Enough?* (New York: W. W. Norton, 1992).

—, "Mobilizing at the Grassroots," *State of the World*, 1989 (New York: W. W. Norton, 1989), pp. 154–73.

Dyson, Freeman, *Weapons and Hope* (New York: Harper and Row, 1984).

Eckholm, Erik P., *Down to Earth: Environment and Human Needs* (New York: W. W. Norton, 1982).

Egginton, Joyce, "Just a Farm Story," [toxic waste incinerators and agriculture] *Audubon*, 87 (November 1985), pp. 134–43.

—, "The Long Island Lesson," [polluting the ground water] *Audubon*, 83 (July 1981), pp. 84–93.

Ehrlich, Paul R., and Anne H. Ehrlich, *The Population Explosion* (New York: Simon and Schuster, 1990).

—, *Extinction* (New York: Random House, 1981).

Eicher, Carl K., "Facing Up to Africa's Food Crisis," *Foreign Affairs*, 61 (Fall 1982), pp. 151–74.

—"Evolving Theory Views Earth as a Living Organism," [Gaia hypothesis] *New York Times*, national edn (August 29, 1989), pp. 17, 19.

Falkenmark, Malin and Carl Widstrand, "Population and Water Resources: A Delicate Balance," *Population Bulletin*, 47 (Washington: Population Reference Bureau, November 1992), pp. 1–36.

Farb, Peter, and George Armelagos, *Consuming Passions: The Anthropology of Eating* (Boston: Houghton Mifflin, 1980).

Farvar, M. Taghi, and John P. Milton, (eds), *The Careless Technology: Ecology and International Development* (Garden City, NY: Natural History Press, 1972).

Freed, Stanley, and Ruth Freed, "One Son Is No Sons," *Natural History*, 94 (January 1985), pp. 10–15.

Freeman, Christopher, and Marie Jahoda (eds), *World Futures: The Great Debate* (New York: Universe Books, 1978).

French, Hilary F., "Restoring the East European and Soviet Environments." In Lester R. Brown et al., *State of the World–1991* (New York: W. W. Norton, 1991), pp. 93–112.

Gasser, Charles S., and Robert T. Fraley, "Transgenic Crops," *Scientific American,* 266 (June 1992), pp. 62–9.

Gibbons, John H., Peter D. Blair and Holly L. Gwin, "Strategies for Energy Use," *Scientific American*, 261 (September 1989), pp. 136–43.

Glantz, Michael H., "Drought in Africa," *Scientific American*, 256 (June 1987), pp. 34–40.

Gray, Andrew, "Indigenous Peoples and the Marketing of the Rain Forest," *The Ecologist*, 20 (November/December 1990), pp. 223–7.

Green, Harold, "The Peculiar Politics of Nuclear Power," *Bulletin of the Atomic Scientists*, 38 (December 1982), pp. 59–65.

Guppy, Nicholas, "Tropical Deforestation: A Global View," *Foreign Affairs*, 62 (Spring 1984), pp. 928–65.

Halle, Louis, "A Hopeful Future for Mankind," *Foreign Affairs*, 58 (Summer 1980), pp. 1129–36.

Hamill, Pete, "Where the Air Was Clear," *Audubon,* 95 (January-February 1993), pp. 38–49.

Hardin, Garrett, "Living on a Lifeboat," *Bioscience*, 24 (October 1974), pp. 561–8.

—, "The Tragedy of the Commons," *Science*, 162 (December 13, 1968), pp. 1243–8.

Holdren, John P., "Nuclear Power and Nuclear Weapons: The Connection is Dangerous," *Bulletin of the Atomic Scientists*, 39 (January 1983), pp. 40–5.

Houghton, J. T., G. J. Jenkins, and J. J. Ephraums (eds) *Climate Change: The IPCC Scientific Assessment* (Cambridge: Cambridge University Press, 1990).

Houghton, Richard A. and George M. Woodwell, "Global Climatic Change," *Scientific American*, 260 (April 1989), pp. 36–44.

Hubbard, Harold M., "The Real Cost of Energy," *Scientific American*, 264 (April 1991), pp. 36–42.

Insel, Barbara, "A World Awash in Grain," *Foreign Affairs*, 63 (Spring 1985), pp. 892–911.

Jackson, Henry, "The African Crisis: Drought and Debt," *Foreign Affairs*, 63 (Summer 1985), pp. 1081–94.

Jones, Philip D. and Tom M. L. Wigley, "Global Warming Trends," *Scientific American*, 263 (August 1990), pp.84–91.

Kennedy, Paul, *Preparing for the Twenty-First Century* (New York: Random House, 1993).

Kramer, Mark, *Three Farms: Making Milk, Meat and Money from the American Soil* (Boston: Little, Brown, 1980).

LaBastille, Anne, "Heaven, Not Hell," [development in the Amazon Basin] *Audubon*, 81 (November 1979), pp. 68–103.

Levine, Adeline, *Love Canal: Science, Politics and People* (Lexington, Mass.: Lexington Books, 1982).

Lewis, H. W., *Technological Risk* (New York: W. W. Norton, 1990).

Lewis, W. Arthur, "The State of Development Theory," *The American Economic Review*, 74 (March 1984), pp. 1–9.

Livi-Bacci, Massimo, *A Concise History of World Population* (Cambridge, Mass.: Blackwell Publishers, 1992).

Lovins, Amory, L. Hunter Lovins, and Leonard Ross, "Nuclear Power and Nuclear Bombs," *Foreign Affairs*, (Summer 1980), pp. 1137–77.

Luoma, Jon, "Gazing into Our Greenhouse Future," *Audubon* (March 1991), pp.53–129.

—, "Trash Can Realities," *Audubon*, (March 1990), pp. 86–97.

—, "Forests Are Dying But Is Acid Rain Really to Blame," *Audubon*, 89 (March 1987), pp. 37–51.

—, "The $33 Billion Misunderstanding," [sewage plants and clean water] *Audubon*, 83 (November 1981), pp. 110–27.

MacNeill, Jim, "Strategies for Sustainable Economic Development," *Scientific American*, 261 (September 1989), pp. 155–65.

Mahler, Halfdan, "People," *Scientific American*, 243 (September 1980), pp. 67–77.

Mason, Jim, and Peter Singer, *Animal Factories* (New York: Crown, 1980).

Maybury-Lewis, David, *Millennium: Tribal Wisdom and the Modern World* (New York: Viking, 1992).

McFalls, Joseph A., Jr, "Population: A Lively Introduction," *Population Bulletin*, 46 (Washington: Population Reference Bureau, October 1991), pp. 1–41.

McNamara, Robert, "Time Bomb or Myth: The Population Problem," *Foreign Affairs*, 62 (Summer 1984), pp. 1107–31.

Meadows, Donella H., Dennis L. Meadows, and Jorgen Randers, *Beyond the Limits: Confronting Global Collapse, Envisioning a Sustainable Future* (Post Mills, Vt.: Chelsea Green Publishing, 1992).

Mellor, John W., and Sarah Gavian, "Famine: Causes, Prevention, and Relief," *Science*, 235 (January 1987), pp. 539–45.

Merrick, Thomas W., "World Population in Transition," *Population Bulletin*, 41 (Washington: Population Reference Bureau, April 1986), pp. 1–51.

Michaelson, Karen L., et al., *Childbirth in America: Anthropological Perspectives* (Westport, Conn.: Bergin and Garvey, 1988).

Miller, G. Tyler, Jr, *Living in the Environment: Principles, Connections, and Solutions*, 8th edn (Belmont, Calif.: Wadsworth, 1994).

Miller, Lynn, *Global Order: Values and Power in International Politics* (Boulder, Colo.: Westview Press, 1985).

Milton, Katharine, "Civilization and Its Discontents," *Natural History* (March 1992), pp. 37–43.

Morone, Joseph G., and Edward J. Woodhouse, *Averting Catastrophe: Strategies for Regulating Risky Technologies* (Berkeley: University of California Press, 1986).

Murphy, Elaine, *World Population: Toward the Next Century*, rev. edn (Washington: Population Reference Bureau, 1985).

Myers, Norman, "The Exhausted Earth," *Foreign Policy*, 42 (Spring 1981), pp. 141–55.

—, *The Primary Source* (New York: W. W. Norton, 1984).

—, "Room in the Ark?" *Bulletin of the Atomic Scientists*, 38 (November 1982), pp. 44–8.

Nafziger, E. Wayne, *The Economics of Developing Countries*, 2nd edn (Englewood Cliffs, NJ Prentice-Hall, 1990).

National Audubon Society, *The Audubon Energy Plan* (New York: National Audubon Society, 1984).

National Resource Council, *Alternative Agriculture* (Washington: National Academy Press, 1989).

—, *Population Growth and Economic Development: Policy Questions* (Washington: National Academy Press, 1986).

—, *Energy in Transition 1985–2010* (San Francisco: W. H. Freeman, 1980).

Nelkin, Dorothy, "Some Social and Political Dimensions of Nuclear Power: Examples from Three Mile Island," *The American Political Science Review*, 75 (March 1981), pp. 132–42.

Newman, Lucile F. (ed.), *Hunger in History: Food Shortage, Poverty, and Deprivation* (Cambridge, Mass.: Blackwell Publishers, 1990).

Pacey, Arnold, *Technology in World Civilization* (Cambridge: MIT Press, 1990).

Pearson, Frederic S., and J. Martin Rochester, *International Relations: The Global Condition in the Late Twentieth Century*, 3rd edn (New York: Random House, 1992).

Peterson, Jeannie (ed.), "Nuclear War: The Aftermath," *Ambio: A Journal of the Human Environment*, 11, no. 2-3 (1982), pp. 76–176.

Population Reference Bureau, *1994 World Population Data Sheet* (Washington: Population Reference Bureau, 1994).

Postman, Neil, *Technopoly: The Surrender of Culture to Technology* (New York: Alfred A. Knopf, 1992).

Prance, Ghillean, "Fruits of the rain forest," *New Scientist* (January 13, 1990), pp. 42–5.

Presidential Commission on World Hunger, *Overcoming World Hunger: The Challenge Ahead* (Washington: Government Printing Office, 1980).

Prosterman, Roy L., *The Decline in Hunger-Related Deaths*, Hunger Project Papers, no. 1 (San Francisco: Hunger Project, 1984).

Raven, Peter, "Tropical Rain Forests: A Global Responsibility," *Natural History*, 90 (February 1981), pp. 28–32.

Reganold, John P., Robert I. Papendick, and James F. Parr, "Sustainable Agriculture," *Scientific American*, 262 (June 1990), pp. 112–20.

247

Repetto, Robert, "Accounting for Environmental Assets," *Scientific American*, 266 (June 1992), pp. 94–100.

—, "Deforestation in the Tropics," *Scientific American*, 262 (April 1990), pp. 36–42.

Rifkin, Jeremy, *Entropy: A New World View* (New York: Viking Press, 1980).

Robey, Bryant, Shea O. Rutstein and Leo Morris, "The Fertility Decline in Developing Countries," *Scientific American*, 269 (December 1993), pp. 60–7.

Romm, Joseph, "Needed–A No-Regrets Energy Policy," *Bulletin of the Atomic Scientists* 47 (July/August 1991), pp. 31–6.

Ross, Marc H., and Daniel Steinmeyer, "Energy for Industry," *Scientific American*, 263 (September 1990), pp. 88–98.

Ross, Philip E., "Hard Words: Trends in Linguistics," *Scientific American*, 264 (April 1991), pp. 138–47.

Royston, Michael, "Making Pollution Prevention Pay," *Harvard Business Review* (November-December 1980), pp. 6–14.

Sant, Roger et al., *Eight Great Energy Myths: The Least-Cost Energy Strategy, 1978–2000* (Arlington, Va.: The Energy Productivity Center of the Mellon Institute, 1981).

Schell, Jonathan, *The Fate of the Earth* (New York: Avon Books, 1982).

Scheper-Hughes, Nancy, "Death Without Weeping," *Natural History* (October 1989), pp. 8–16.

Schumacher, E. F., *Small Is Beautiful: Economics as if People Mattered* (New York: Harper and Row, 1973).

Short, R. V., "Breast Feeding," *Scientific American*, 250 (April 1984), pp. 35–41.

Shulman, Seth, *The Threat at Home: The Toxic Legacy of the US Military* (Boston: Beacon Press, 1992).

Simon, Julian, *The Ultimate Resource* (Princeton: Princeton University Press, 1981).

Simon, Julian, and Herman Kahn, (eds), *The Resourceful Earth* (Oxford, England: Blackwell Publishers, 1984).

Singer, Peter, *Animal Liberation: A New Ethics for Our Treatment of Animals* (New York: New York Review, 1975).

Singer, S. Fred, "Warming Theories Need Warning Label," *Bulletin of the Atomic Scientists*, 48 (June 1992), pp. 34–9.

Sivard, Ruth, *World Military and Social Expenditures 1993*, 15th edn (Washington: World Priorities, 1993).

Smil, Vaclav, *China's Environmental Crisis: An Inquiry into the Limits of National Development* (Armonk, NY: M. E. Sharpe, 1993).

—, *Energy in China's Modernization: Advances and Limitations* (Armonk, NY: M. E. Sharpe, 1988).

—, "Ecological Mismanagement in China," *Bulletin of the Atomic Scientists*, 38 (October 1982), pp. 18–23.

Spinrad, Bernard, "Nuclear Power and Nuclear Weapons: The Connection is Tenuous," *Bulletin of the Atomic Scientists*, 39 (February 1983), pp. 42–7.

Sprout, Harold, and Margaret Sprout, *The Context of Environmental Politics* (Lexington: University Press of Kentucky, 1978).

Stanislaw, Joseph, and Daniel Yergin, "Oil: Reopening the Door," *Foreign Affairs*, 72 (September/October 1993), pp. 81–93.

Steinhart, Peter, "Down in the Dumps," [solid wastes] *Audubon*, 88 (May 1986), pp. 104–9.

Stenehjem, Michele, "Indecent Exposure," [contamination from nuclear bomb plant] *Natural History* (September 1990), pp. 6–21.

Stobaugh, Robert, and Daniel Yergin (eds), *Energy Future: Report of the Energy Project at the Harvard Business School* (New York: Ballantine Books, 1980).

Teitel, Martin, *Rain Forest in Your Kitchen: The Hidden Connection Between Extinction and Your Supermarket* (Washington: Island Press, 1992).

Teitelbaum, Michael S., "The Population Threat," *Foreign Affairs*, (Winter 1992/1993), pp. 63–78.

Tien, H. Yuan, et al., "China's Demographic Dilemmas," *Population Bulletin*, 47 (Washington: Population Reference Bureau, June 1992), pp. 1–44.

Tierney, John, "Betting the Planet," *New York Times Magazine*, (December 2, 1990), pp. 52–81.

Toon, Owen B., and Richard P. Turco, "Polar Stratospheric Clouds and Ozone Depletion," *Scientific American*, 264 (June 1991), pp. 68–74.

Uhl, Christopher, "You Can Keep a Good Forest Down," *Natural History*, 92 (April 1983), pp. 71–9.

United Nations Development Programme, *Human Development Report 1993* (New York: Oxford University Press, 1993).

United Nations Environmental Programme, *Environmental Data Report* (Oxford: Blackwell Publishers, 1991).

US Environmental Protection Agency, *Progress in Groundwater Protection and Restoration* (Washington: US Environmental Protection Agency, 1990).

Washington, Warren M., "Where's the Heat?" *Natural History* (March 1990), pp. 67–72.

Weaver, James, and Kenneth Jameson, *Economic Development: Competing Paradigms* (Washington: University Press of America, 1981).

Weinberg, Alvin M., "Deterrence, Defense, and the Sanctification of Hiroshima," *The World and I* (June 1989), pp. 467–91.

Weintraub, Pamela, "The Coming of the High-Tech Harvest," *Audubon* 94 (July–August 1992), pp. 92–103.

Weir, David, and Mark Schapiro, *Circle of Poison* (San Francisco: Institute for Food and Development Policy, 1981).

White, Robert M., "The Great Climate Debate," *Scientific American*, 263 (July 1990), pp. 36–43.

Whitehead, Barbara Dafoe, "Dan Quayle Was Right," [divorce harms children] *The Atlantic* (April 1993), pp. 47–84.

Williams, Ted, "Hard News on 'Soft' Pesticides," *Audubon*, 95 (March-April 1993), pp. 30–40.

—, "The Metamorphosis of Keep America Beautiful," *Audubon* (March 1990), pp. 124–34.

Wilson, Edward O., *The Diversity of Life* (Cambridge: Harvard University Press, 1992).

Winner, Langdon, *The Whale and the Reactor: A Search for Limits in an Age of High Technology* (Chicago: University of Chicago Press, 1986).

Wolfe, Alan, *Whose Keeper? Social Science and Moral Obligation* (Berkeley: University of California Press, 1989).

World Bank, *World Development Report 1992 – Development and the Environment* (New York: Oxford University Press, 1992).

—, *World Development Report 1990 – Poverty* (New York: Oxford University Press, 1990).

World Commission on Environment and Development, *Our Common Future* (New York: Oxford University Press, 1987).

World Health Organization (WHO), *Public Health Impact of Pesticides Used in Agriculture* (Geneva, Switzerland: WHO, 1990).

World Resources Institute, *World Resources 1994–95* (New York: Oxford University Press, 1994).

—, *The 1993 Information Please Environmental Almanac* (Boston: Houghton Mifflin, 1993).

Yergin, Daniel, The Prize: The Epic Quest for Oil, Money, and Power (New York: Simon and Schuster, 1990).

Index

251